C0-AKP-950

THE AWKWARD SPACES OF FATHERING

For the fathers who appear in this book
and for the countless others who are part of my coming community

The Awkward Spaces of Fathering

STUART C. AITKEN
San Diego State University, USA
and The Norwegian University of Science and Technology

ASHGATE

Published by
Ashgate Publishing Limited
Wey Court East
Union Road
Farnham
Surrey, GU9 7PT
England

Ashgate Publishing Company
Suite 420
101 Cherry Street
Burlington
VT 05401-4405
USA

www.ashgate.com

British Library Cataloguing in Publication Data
Aitken, Stuart C.
The awkward spaces of fathering
1. Fatherhood
I. Title
306.8'742

Library of Congress Cataloging-in-Publication Data
Aitken, Stuart C.
The awkward spaces of fathering / by Stuart C. Aitken.
p. cm.
Includes bibliographical references and index.
ISBN 978-0-7546-7005-6
1. Fathers. 2. Personal space. 3. Interpersonal relations. I. Title.

HQ756.A38 2009
306.874'2--dc22

2008045723

ISBN: 978-0-7546-7005-6 (Hb)
ISBN: 978-0-7546-8821-1 (eBook)

Printed and bound in Great Britain by
MPG Books Ltd, Bodmin, Cornwall.

Contents

PART II CLOSING IN

PART III MOVING

PART IV STOPPING

List of Figures

Preface

The wife of one of the fathers in this book collared me at a community meeting saying, "I want to talk to you about that chapter you wrote. Billy let me read it." I tensed, my head raced towards a plausible defence against whatever she was about to say. I was on the cusp of sending the manuscript to my editor at Ashgate and I did not want to make more changes. The fathers whose stories pepper the pages of this book were all afforded an opportunity to comment on draft chapters and we excised some of the material that was troubling. I hadn't thought about giving their partners the same opportunity. Billy's wife's eyes glassed over as she stared at me. Rage? I really didn't want to deal with an irate partner.

"Shit, Stuart," she said. "We read it together and we cried. We cried together. You nailed Billy. That is exactly him. And you nailed his relationship with his son. I cried when I read the poetry."

My heart stopped racing. Wow! This, I thought, is exactly what I intended when I started this project. I wanted to do more than represent the spaces of fathering, I wanted to find forms that express the work of fathers as an emotional practice. Certainly, I wanted to talk about the complex and incomplete histories and geographies of fathering and I wanted to elaborate theories of space through the lives of men and women who are fathers, but I also wanted to touch an emotional core because that, I believe, is the heart of my academic practice and it is the heart of my fathering. To the extent that this book is about my fathering, it is also about my poetics of becoming something different through my connection with these men and their families.

Given my penchant to celebrate the emotional work of fathering, I am still surprised by how many men who appear in this book feel profound emotions at re-reading their stories. Part of their charge is recognition of how things change. None of the fathers in this book is where he was when we sat down for a discussion about his fathering. Things change. One of the fathers is now separated from his partner and is struggling to maintain a connection to his son; another is grieving his daughters' seeming rejection of his desire to spend more time with them. Several are more connected than ever to their children and are working hard at co-creating communities that provide emotional safety and security. One is thinking about ways of enfolding more children into his communal household. Another has secured federal funding to help improve his community's infrastructure, including the establishment of youth programs. Things change. Men care and take responsibility. This book reflects, in some small and yet important way, the heart of that care and responsibility.

A father whose story appears in this book famously quipped that "[t]hese are the times to try men's souls" (Thomas Paine, 1776/1979, 55). In these times of continued war, threats of terrorism, rampant neo-conservatism, growing economic and financial insecurities, the disintegration of habeas corpus and other civil liberties, and neo-liberal pressures on families to shoulder an increased burden of responsibility for health, education, housing and other material contexts that were once secured by the state, I am nonetheless hopeful when I sit with men and they share with me what they consider important. If 'habeas corpus' is literally about 'delivering the body' then the irony of its erosion as a right of due process in the US and the UK is that there is, I believe, a growing embodiment of caring and responsibility that is storied and powerful, that is about the day-to-day emotional work of fathers, mothers, children, aunts, uncles, and strangers who come together in communities that are emergent, transcendent, inclusive and loving. This book is a celebration of that day-to-day emotional work.

Stuart C. Aitken
La Mesa, California, February 2009

Acknowledgements

I want to thank the numerous people who supported my efforts on the 'awkward fathering' project these last ten years. Peg, Ross, Catherine and (yes) Maggie never stopped believing in me or the project. The men, women and children who are part of my coming community deserve a special mention. Here are only some of the names (you know who you are): Jim, Brian, Sharon, Mark, Connor, Tim, Sam, Ed, Serge, Janet, Savannah, Gerry, Sue, Anna, Anne, Cheryl, Jodie, Teresa, Evan, Ron, Rob, Ken, Kenny, Michael, Heidi, John, Fernando, Fernanda, Tom, T, Allen, Dan, Patti, Diana, Scott, Tina, Tiana, Melissa, James, Sean, Elaine, Sally, Eddie, Raul, Bob, Mike, Silvia, Reg, Hannah, Debbie, Kirsten, Christopher, Hamish, Rory, Laura, Jo, Mary-Ellen, Allan, Christie, Felipe, Kathy, Stuart, Steve, Clem, Lilia, Robert, Andrew, Vicky, Dave, Giorgio, Ragnhild, Bruce, Norman, Chris, Ernie, David, Morna, Joel, Alasdair, Maggie, JP, Margaret, Aage, Bill, Cindy, Robin, Cindi, Will, Catherine, Doug, George, Guy, Bobby, Terry, Joan, Larry, Sandy, Keith, Gary, Phil, Greg, Bonnie, Jan, Polly, Maurizio, Derek, Lynn, Calum, Keith, …

I also want to thank the institutional and academic support of San Diego State University, the Norwegian University of Science and Technology and the University of Wales. Amongst other things, many of the faculty, staff and students of these institutions joined me (with seminars, workshops and comments) in a celebration of this work as it progressed to the book and as it continues. I want to single out, in particular, Chris Moreno and Giorgio Curti who read a draft of the manuscript and provided invaluable comments. Our work continues.

Introduction

He carved the word fuck
... into his leg.
It was all part of him.
He moved in and out,
psychiatrics and psychologists.
He'd just sit there like this
... with his legs folded.
The biggest, the hardest thing I ever did
was leave him,
when we had to leave him,
at the hospital
when he was cutting himself.
That was the fucking toughest thing I ever done.
He was maybe 14.
That is when I discovered how old I was.

(Billy, November 11, 2007)

Billy knows that fathering is hard work. He knows the work is an ill-defined and ongoing emotional process. He knows that it is full of surprises, and those surprises open up for Billy opportunities to connect more fully with his sons, or not. There is choice. Sometimes he chooses to connect and sometimes he feels too old and too tired. At those times, the work of fathering drains emotionally and physically.

The words in the epigram are derived from the transcription of a long conversation at Cosmos' Café in La Mesa, California. I call it ethnopoetry: a poetry that simultaneously refracts a moment of conversation, recreates a connection between two people and reconnects a series of previous encounters. The words are pulled apart, sutured, contracted and re-worked into fifteen lines contrived poetically as a reflection of Billy's fathering as I see it and as described to me over the last several years. The poem expresses some considerable anger; it is an anger that covers a profound sadness. The anger and sadness are part of my fathering also; my connection with Billy enables the ethnopoetry. At this moment, Billy is tired. He is remembering a conflict with his oldest son that ended in hospitalization. The toughest thing he ever did. He is remembering the frustration, the doctors, the constant monitoring of his son's health. He is remembering the shame and the heartache. In the course of our conversation, his voice begins to tremble and his eyes glass over, and the sad face of a needy child appears, barely discernible behind the tears. I want to re-assure the child, to tell him everything will be alright. I want to be the father who wasn't there for young Billy.

Billy and I are drinking tea. We sit just off the pavement in the café's outside patio. The afternoon is warming. A light breeze connects together bamboo chimes suspended in the branches of a cicada tree. The tree umbrellas our patio and a large portion of the pavement beyond. A humming-bird flits doggedly between several small red flowers. A cat rubs against the tree. My dog, Maggie, licks Billy's hand. Someone told me that Labradors pick up on the emotions around them, and that is why they are ideal therapy-dogs. Another tear rolls down Billy's cheek. He tells me how he is much more responsive to emotions today. How, in this way, he is different from his father.

"My dad was a very, very macho man," he says.

Billy's dad worked as a sheriff in a small town. He tells me of the day that a man stood in the centre of the street holding a loaded shotgun. Pedestrians scattered. Billy's dad walked into the street. I imagine a bow-legged man with a barrel chest, there is a sheriff's star pinned to his chest. A cowboy hat is tilted slightly off his brow. The man's got Billy's sad blue eyes. Those eyes are staring down the man holding the shotgun. It is a classic Western stand-off.

"He walked right up to the guy," says Billy, "took the shotgun out of his hands. It was just … you know?"

My image of the barrel chest and the sheriff's star evaporate. It did not work. What image fits? What did I know of the image of the father? I am not sure that I did know at the time, but now, as I remember my afternoon with Billy in the quiet of my office – a different time, a different place, a different task – I think that perhaps I know something. The sad blue eyes remain. I know the emotions – the anger, the sadness, the questioning. They are part of the work of fathering. Perhaps I can construct the image of the remembered father. Felt then, recollected now, pointing towards a something different: a compass rose of memory. The macho cowboy did not so much evaporate as fade into the background. His presence remains.

What the compass rose points to are fathers today contextualized differently from previous generations, as well as fathers contextualizing themselves in different ways. These are leanings into, and trajectories through, men's lives. I call this emotional work *fathering*. There are also institutional tracings of men's lives that I call *fatherhood*. This book is about those leanings, trajectories, works and tracings. It is about how men do fathering today, and how that work is different from traditional constructions of fatherhood. It is about change and transformation, and it is also about what is learnt and what is carried through. What comes together in the pages that follow is a rendering of fathers – their works and their emotions – that is partial and only vaguely coherent. What are the doings – that is, the emotional works – that create fathering? The question suggests a connection that is both fleeting and dogged. It is a connection from which I try to produce a mapping that is simultaneously a space of meditation, a space of practice and a space of surprise.

Put another way, the book is an attempt to uncover what Michel de Certeau (1984) calls "spatial stories" that *reflect* the *everyday mappings* that are men's lives and their practices as fathers, and it is about everyday spatial stories that *refract* the *tracings* that are ideas of fatherhood. Mapping fathering as a celebration – as I try to do in this book – is to embrace this awkwardness and incoherency.

The Awkward Spaces of Fathering

The book elaborates the awkward spaces of fathering through diverse geographical stories. These stories include a variety of historical transformations although the focus of the book is fathering over the last 20 or so years. Along the way I try to recover fatherhood – partial and incomplete – as both idea and action. I spend some time with ideas of fatherhood from the Enlightenment onwards but, by so doing, it is not my intent to set a stage or to uncover any kind of coherent notion of fatherhood. Rather, I want to note some of its transformations with a very particular focus on fathering histories and geographies as awkward and incoherent. The book's main focus is contemporary fathering within the changing spatial context of Anglo-American society. The bulk of the book comprises conversations with fathers who live and practice their fathering alongside what is famously characterized as today's crisis of masculinity.[1]

Of course, there are always – and there always should be – crises around what constitutes identity. At least to some degree, the awkward spaces of fathers' lives are a psychotic surrender; both a *respite* and a *panic* before the hysteria that emanates from crisis. What I mean by this is that, from an ontological perspective that eschews structure, any coherent form of *the* father necessarily melts away and fades as soon as it begins to find substance. And so, a move towards hysteria – an existential crisis, if you will – ensues prior to another bout of *mauvais feux* – the bad faith that Jean-Paul Sartre (no date, 47) warns us against – and returns me to another potential framing of fatherhood as an ongoing process. The emotions embodied by respite and panic intrigue me, because it is a surfeit of emotions that moves me beyond framings to new practices and new potentialities. William Connolly (1999, 146) notes that a recognition of this is part of an active politics of becoming, although he cautions that new potentialities can easily become reintegrated into old framings – "old piles of argument" – rather than new ways of being. Herein lies an important spatiality.

1 Bob Connell (1995, 83–4) was the first to document tendencies towards what he calls an "ontoformative gender crisis" in which a variety masculinities and femininities exist in problematic "configurations of gender practice." He points out that transformations in power relations within the family show the most visible evidence of crisis tendencies with the historic collapse of the legitimacy of patriarchy, with the global movement to emancipate women and with a move toward economic equality between the sexes. Connell says very little about the differentiated responses of fathers to the crisis of masculinity, and so my project is to elaborate that particular affect through the emotional work of fathering.

In practice, the crises of masculinities both foment and are created by the awkward spaces within which fathers live. These are the spaces that define the relationship between men, women and children; between men and other men; between men and families and men and communities; between fathering as a practice and fatherhood as an idea. The spaces are awkward because defining the context of embodied practices is never completely comfortable. Recognition of the awkward spaces we live in and produce define who we are and what we do: they are part action, part illness, part remedy, part us, part them, part it, part fantasy, part fact.

Fathering Fantasies

The ideas of place, space and fathering that launched my thinking about the awkward spaces of fathering began in the 1990s with a project entitled *Family Fantasies and Community Space*. The project connected with 577 adult members of households in San Diego who were expecting a first child. I followed about one-third of those parents with yearly in-depth interviews for the first four years of the child's life. I was interested in how a first child changed gender performances within households; I explored the ways that the day-to-day spaces and practices of mothering and fathering were transformed in the early years of child-rearing (Aitken 1998). As part of this project, I explored the rites and practices that surrounded childbearing (Aitken 1999), changes in gender roles and relations between parents and how those impacted children (Aitken 2000a), and the ways environments fostered particular forms of childrearing (2000b). I was also interested in the notion of the nurturing father; of uncovering stories that the academic and popular media of the time were dubbing 'the domesticated dad' (Aitken 2000c, 2005).

At the same time, I was becoming a father. Ross was born in 1990 and Catherine in 1992. Like many fathers today, I was present for the births and I worked at being around as much as possible during their first years. Just tenured at San Diego State University, and with the aid of colleagues who picked up my classes, I was in the enviable position of taking some time away from work. Peg, my partner in all of this, gave up her teaching job while the children were infants.

Some weeks after he was born, I remember fussing around Ross and a change of diapers one afternoon. I was back teaching my classes and I was at home as much as possible. Peg burst into the kitchen in a flurry of post-partum anger.

"What are you doing here?" Peg charged. She may have used more colourful language as she was clearly enraged. "Shouldn't you be at work? Don't you have a contract with the State of California? Don't you have responsibilities elsewhere?"

First the knife thrust and then the twist.

"You don't know what you are doing here. You are not helping. Go to work!"

At that point my world changed.

Peg and I laugh at those times now, but we do so with caution. Neither of us knew what we were doing. They were trying times and we were helped through them only with important emotional connections and support from a variety of institutions and a growing community of friends and acquaintances. Out of those times and continuing today we grow and change as parents, partners and as members of larger communities.

The *Family Fantasies* project suggested that at the level of day-to-day fathering practices, close attention should be paid to the distinction between men 'helping out' and men taking responsibility for childcare. On looking at fathers' behaviours more closely, it seemed to me that the research on nurturing men was misplaced because it did not account for the space and work of fathering in relation to patriarchal norms that seemingly continued to define fatherhood in spite of Connell's (1995) proclamation that the legitimacy of patriarchy had collapsed. From what came out of *Family Fantasies,* and from some of my own fathering experiences, I was unable to reject the notion that "fathers embrace the emotions of parenting without taking on the responsibility for domestic labour that such emotions should entail" (Aitken 2000c, 597). This concerned me greatly, and was a large part of the impetus for the *Awkward Spaces* project.

Fathering Facts

Popular media and empirically-based social science studies often elide larger issues of men's responsibilities by continuously creating and reifying fatherhood as an evolving set of practices that link to other important familial changes. All the changes point, the evidence suggests, towards fathering as a form of co-parenting.

Associated Press reporter David Crary (2008) summarizes a number of research articles on fathering practices under the title "Increasingly, Men Pitch in on Housework." According to two social science research studies discussed by Crary, men's contribution to housework doubled over the last four decades and the time spent on child care tripled over that same time period. This is good news, notes Crary (2008, A7), although it does not imply equality between men and women over domestic chores. Rather, it suggests "that the rules of the game have been profoundly and irreversibly changed." The data are incontrovertible, he says. The change is here and now. Time budget studies by Scott Coltrane and Oriel Sullivan (cf. Coltrane 1996; Sullivan 2004) show how men in the United States and Israel tripled the amount of time they spent on childcare between 1965 and 2003. Coltrane and Sullivan predict that men's contributions will increase further as more women take jobs. This, of course, relates to larger changes in women's paid employment:

> Men share more family work if female partners are employed more hours, earn more money and have spent more years in education (Sullivan and Coltrane, cited in Crary 2008, A7).

Reflecting on these changes, Wendy Manning and Pamela Schmock (cf. Manning and Schmock 2005) point out that a persistent gender gap remains for what they call "invisible" household work such as scheduling medical appointments, buying gifts and organizing birthday parties. These kinds of activities are not captured in time-budget diaries, nor are connections to responsibility and emotional engagement. This is the kind of responsibility gap to which the *Family Fantasies* project pointed: what, precisely, is a 'child-caring' man?

"Child-caring men," I argue in an article that focuses on men who take primary responsibility for rearing their children, "are disembodied father-figures because they cannot take on the work and responsibility of parenting without imagining themselves as 'Mr. Mom'" (Aitken 2000, 597). From this I was left with a series of questions that my discussions with fathers in *Family Fantasies* could not address. These questions revolve around the emotional work of fathering as it is manifest in families and communities, and the ways fathering evolves from families and communities.[2] How do fathers define themselves in ways that are about fathering rather than mothering? With the supposed dissolution of patriarchy, what is equal about the communal household (fathers and mothers as partners)? If there is an emotional heart to the communal household, where does it reside? What are the relations between fathers and the co-constructed spaces of the family and household? What would a community of fathers look like? For the most part, these questions concern the ways fathers' emotions shape their spaces of connection – sometimes stratified, sometimes awkward, sometimes colluded, sometimes smooth – with families and communities. There are a myriad of ways these questions may be addressed.

Modes of Encounter

As part of a poststructural package that includes the work of Foucault, Derrida, Deleuze and Guatarri as well as Connolly, Massey, Butler and Gibson-Graham, non-representational and relational theories have garnered some currency in geography and other social sciences because they elaborate a less structured way of knowing that is not about hierarchies (family, community, state) or dichotomies (mother/father). They get to issues of how things work (what they do) by elaborating multiple modes of encounter – and especially multiple emotional encounters – that are not about fixing identity through representations but, rather, are about opening up identities to a politics of hope.

I began the *Awkward Spaces* project as a long-term endeavour to get a sense of fathers' emotional and relational connections to families and communities. I

2 My discussions with mothers, alternatively, enabled a close look at the ways their parenting practices connected with family and community so that I could say something about the joint problematics of community scales and the emotional push of mothering (Aitken 2000d).

connected with a variety of men's groups so that I could listen to the stories of fathers.[3] My focus on emotions continues from the *Family Fantasies* project and is tied to a set of evolving theories in the social sciences that point to new ways that events and practices are encountered and understood. I spend some time in the chapters that follow elaborating this theoretical connection to the practical and emotional work of fathering. In so doing, I begin to get a sense – halting and incomplete – of the ways hierarchies dissolve and make little sense when fathers' emotional connections bring together families, places and communities.

As a pedagogical device, I divide the book into four main sections, although the fathering stories slip between these somewhat arbitrary designations. Part I focuses on FRAMING to the extent that it highlights the institutional constructions of fatherhood and larger contexts that spatially frame the subject of *the* father. The emphasis is on spatial frames as a contrivance for potential de-territorialization and re-territorialization. Part II, entitled CLOSING IN, comprises three chapters that highlight the 'geography closest in': bodies and emotions. By coming this close, the section also broaches ethical dilemmas around the natural rights of fathers. The MOVING of Part III is about tactics, strategies, migrations and lines of flight, and the STOPPING in Part IV is about the impossibility of sedentarism and domesticity as fathers contextualize home.

In Chapter 1 I introduce my methodological leanings and begin my modes of encounter. I use anecdotes from the stories of two fathers – Ed and John – to talk about the importance of memory, of stories, of hearth and home, and of families and communities. I offer a series of lines of flight from, back to, and then, again, away from fathering as an event in space (Deleuze and Guattari 1987; Stohmayer 1998). My stories are circuitous and flighty, but they are not without important connections and groundings. For the most part the connections are theoretical but some are empirical. For example, the story of hearth and home that begins with John in Chapter 1 – and contains a Heideggerian notion of being-at-home-in-the world – returns in Chapter 11 when Tony (an African-American father whose home comprises three generations) discusses home as the ultimate expression of openness. John's brief introductory example provides a book-end to Tony's story, and between these stories I take flight with fathers in exile, transnational fathers, fathers journeying outwards and inwards, spiralling fathers, fathers telling stories without beginnings or ends. Through their doings (journeys, exiles, home-makings, nurturings), these men create fathering.

3 From 1998 onwards I connected with social workers and professionals in California's CPS (Child Protective Services). I also joined a variety of men's circles and therapy groups and participated in men-only groups of Alcoholics Anonymous. The initial intent of the *Awkward Spaces* project was to collect stories on men who were separated from their families (by court order or by request of family members) and to document the ways they tried to re-connect. Over the years, I got to know these and other fathers well enough to broaden the project to create an exegesis of the emotional and awkward spaces of fathering.

My first encounter with a framework for fathering is with John. I use his story in Chapter 1 to question the possibility of spatial frames. I continue this line of thinking in Chapter 2, where I trace Enlightenment thinking about fathers and patriarchy while at the same time mapping some different, less well known, fathering trajectories. In particular, I highlight the fathering practices of a couple of Enlightenment figures. I take a bold stance here, arguing that a focus on Enlightenment empiricism and logic suggests a particular form of distanced fathering whereas a closer look at some of the writers of this time – their writings and their doings – suggests the importance of emotions and commitment. In Chapter 3 I broach the notions of difference and the latent emotion of fathering through the stories of Ed and Benjamin. Chapter 4 is the first of two chapters where I look at movie representations of fatherhood. These moving images are also partial stories of fathering. My intent is to punctuate the stories from my ethnographies with these thumbnail (small visual sketches) representations. In Chapter 4 I focus on movies that focus on fathering as an embodied practice. I do so, in part, to provide visual relief from the ethnographies, but also to render some quirky visual transgressions of *the* father: a different mode of encounter. Ethnography and film provide two modes of encounter that mix representational and non-representational ways of knowing in an attempt to raise fathering as a work, a doing, a toil that is simultaneously what I know (and do not feel) and what I feel (and do not know).

In Chapter 5, I return to the ethnographies, with poetry and dialogue around the inevitability of fatherhood. As an adoptive parent, Andy helps me paint a picture of fatherhood as a seemingly natural event. This theme is continued in Chapter 6, where Cindy performs a very specific form of fathering and calls into question – in a different way – the naturalness of fatherhood. Chapters 5 and 6 point to the ways fathering is constructed as an emotional and practical work, rather than a natural right. Chapter 7 is the other movie interlude. With this thumbnail visual sketch I focus on fathers in exile. This is followed in Chapter 8 with the story of Quixote, a migrant labourer from Mexico and the evolving truth of his journey into exile and back towards fathering. With Chapter 9 I begin to come home, with the recognition that being-at-home is also an important series of movements. I talk about the trouble with space and use Rex's geographic solutions to solve his relationships with his sons to elaborate aspects of that trouble. This – along with the stories of Fred and Stan – enables a discussion of domesticity in Chapter 10, which sets me up for coming home, and Tony's heroic generational saga in Chapter 11.

Each chapter in this book encounters fathering in a different way. Although there are many connections between the stories, there are four experiments – two that are methodological and two that theoretical – with which I characterize the book. The experiments are movements or trajectories, but that are also pretexts of larger arguments that I want to make as the book proceeds, so they are also connected one with another. First, as I've already intimated, I want to experiment methodologically with new ways (at least for me as a social scientist) of communicating the contexts of ethnographies. I try to use a mixture of style that brings my participants more fully into the text, and intersperses their stories, thoughts, the poetry of their words and their silences, with nuggets gleaned from the academic literature. It seems to me that past writing using interviews and ethnography (and especially my own) communicates in exactly the opposite way: a slurry of academic theory and scholarly argument interspersed with a few choice nuggets from interviewees. What I want to do here is 'cherry pick' from theory to illustrate the stories from the *Awkward Spaces* project rather than the other way around.

My hope is that my switching from data to theory, from interview dialogues to ethnopoetry to academic literatures, makes sense. At the very least, what I want to do here is move beyond the somewhat dry rendering of interview transcripts in the form of grey columns of quotations, to engage in the emotional contexts of the interviews and ethnographies, to represent as best I can the tears and laughter that are part (a very large part) of my connection with these men. In discussing the place of emotions in research, Liz Bondi (2005) calls for this kind of rendering in our writing. If this book is about the emotional work of fathering, it is also about "the ubiquity of emotion work" (Bondi 2005, 232) within social science research.

A second methodological point, which is clearly related to what I just said, is that emotions are both private and co-constructed. My conversations with the fathers in this book are charged around the context of their and my fathering.[4] As you read this text you are no doubt engaged in a series of emotions about which I am quite anxious. Does the text enable you to connect with these fathers? Are you intrigued? Do you want to read on? Are you bored? Is it time to move on to a different text by a more interesting writer? So this is not just about my last several years of engagement with fathers, it is also about how I communicate that engagement with you. It is about today: this moment in time and this place.

Third, and here I turn to theory, I want to experiment with the contemporary theoretical category of *the* father in a number of ways. This requires in the chapters that follow a rendering of the known histories and geographies of the creation of fatherhood as an idea, and its institutionalization through various machinations of patriarchy. And yet the notion of patriarchy in and of itself raises some interesting denouements in fathers' stories, in the reality of fathering practices, and in the emotional work of fathering. It turns out, and here I presage one of my most important conclusions, that the work of fathering – at least with the men I engage – hardly ever is about the authority of the father. So where precisely is patriarchy located? What happened to the spaces of patriarchy – to the law of the father – that historians so clearly delineated through the 19th and 20th centuries? To a degree I take this up in Chapter 2, and here I am solaced by Liz Wilson's (1993) admonition

4 A large part of the work that comprises this book came at a time when I struggled to accept my tendencies towards alcoholism after the death of my father and, importantly, as I saw the ways my drinking impacted my relations with Peg, Ross and Catherine. What remains private in the pages that follow are some of the events in my life and in the events of the lives of the other fathers that might cause pain to family members and those close to us.

that transgression by definition requires a category and so the patriarchal category of the "law of the father" is a bit of a straw dog that I raise on occasion and then dispense with as close-to-useless. My hope is that the concept does not straight-jacket my arguments.

My fourth experiment – and this point is perhaps most critically related to the kinds of transformations I am trying to get at with both the movies and my conversations with fathers – pushes the questions: why fathers and why now? How do we come to terms with the practice of fathering through the larger theoretical constructions that permeate both neo-liberal and post-capitalist thinking? Sallie Marston and her colleagues (2005) call for a flat ontology, which recognizes that dependencies today are most often relational rather than hierarchical. Indeed, the construction of some hierarchical ontology that reaches from the body through the family (the father?) to the community (the patriarch?), to the nation (the fatherland?), to the globe and beyond (one giant step for mankind?) is increasing difficult to conceptualize. The work of the Gibson-Graham collective (1996, 2006) is very useful here: I try to move with their notion of *communal households* as part of what might be thought of as a kind of gathering, "a non-kin-based family that *is* a community economy" (Gibson-Graham 2006, xii). I also move beside Giorgio Agamben's (1993) notion of the *coming community* as something theoretically "out of bounds, uncontained by the disciplines, insubordinate … practices of resistance … inventing, excessively, in the between … processes of hybridization."

In the pages that follow I collect together stories of fathers who represent very different positions in terms of income, ethnicity, sexual-orientation and circumstance. At one extreme I have fathers who are ordered by the court system not to see or visit their children and at another I have fathers who struggle for years to conceive children with their partners; in poetry, dialogue and thumbnail visual sketches, the stories range from fathers who arrived in the United States as illegal migrant labourers to middle-class white collar workers in suburbia; I raise issues for lesbian women who raise sons by taking on the role of a father, and fathers who deal with children who 'come out' with a different sexual orientation. In these stories, a common theme revolves around struggles to maintain an amorphous and yet critical *communal household/family*. Whereas elsewhere I talk about the shared lives of fathers and mothers (Aitken 1998, 2001a), mothers (Aitken 2000d) or children (Aitken 1994, 2001b), this book is devoted to fathers. My choice is guided in part by stories of fathers who fight hard to maintain or regain some sense of parenthood, a sense of family and a sense of community. There is nothing in this book that is generalizable, naturalizable or essentializable into a category 'father'. Fathers' struggles comprise a journey, a uniquely spatial series of movements towards an ill-defined goal. Often, the stories of these journeys remain hidden behind notions of masculinity and fathering that forefront more public forms of identity but when excised from – and yet uniquely contextualized beside – these contexts through expressions of emotion there is evidence of a clear work. Ultimately, with halting steps, the book moves to the opening of possibilities of becoming other.

The Poetics of Becoming Other

> We have found that we need technologies for a more reticent yet also more ebullient practice of theorizing, one that tolerates 'not knowing' and allows for contingent connection and the hiddenness of unfolding; one at the same time foregrounds specificity, divergence, incoherence, surplus possibility, the requisite conditions of a less predictable and more productive politics (Gibson-Graham 1996, xxxi).

Awkward spaces are favoured as a territory for 'becoming other'. This territory speaks to the problematic distinction between 'being' and 'becoming' that warps discussions of subjectivity and focuses them into simplistic dichotomies. It is difficult to embrace the 'not knowing' and 'hiddenness' of the unfolding implied in becoming and so, a default, 'becoming-the-same' continues as a painful recapitulation of norms, while 'being' carries with it a huge burden of essentialism and separateness. Alternatively, Brian Massumi's (1992, 95) use of the concept of 'becoming other' invites, in a post-structural Deleuzian sense, "each contained and self-satisfied identity to be grasped outside its habitual patterns of action, from the point of view of its potential, as what it is not, and has never been, rather than what it has come to be." In more spatially explicit terms, the concept draws on Deleuze and Guattari's (1987) notions of deterritorialization and reterritorialization, where the former process deconstructs the individual and thus opens up new possibilities for existence, while the latter reassembles this intensity to form a new identity that is viable within the context in which it finds itself.

Massumi (1992, 102) goes on to argue the possibilities of becoming other as a group project. Embracing this perspective, Gibson-Graham (2006) see the wisdom of Deleuze and Guattari's theorizing as a generative ontological, community-oriented centripetal force working at the pull of essence and identity towards difference and differentiation. Using the technique of what they call 're-reading', they argue for the uncovering of what is possible but obscured from view. Similarly, Bob Connell (1995) argues that the dominance of psychoanalytic interpretative strategies in men's studies has occluded questions about technique. He argues for an understanding of men in which interpretation remains, and is coupled with a focus on techniques that foreground the affective aspects of men's experiences.

An affective re-reading from contingency (as I am trying to do here) rather than necessity situates essentialized, naturalized and universalized forms – like the father, the family, the community, the market, the self-interested and self-actualized subject – in specific geographic and historic locations, releasing them from an ontology of structure or essence. The Gibson-Graham (2006, xxxii) mandate to "reread for difference rather than dominance," at least as a starting place, is undergirded by Deleuze and Guattari's theorizing.

Gibson-Graham (1996, 82) note the ways Deleuze and Guattari challenge the metaphysics of presence in Western post-Enlightenment thought. Rather than positioning identity in a fixed, grounded, Cartesian grid, a Guattarian space evokes air, smoothness and openness while a Deleuzian subjectivity focuses on potentialities. With this way of thinking, note Gibson-Graham, identity is splintered into the not-knowing of multiplicity, heterogeneity, rupture and flight. It is a map, not a tracing.

> The map is open and connectable in all of its dimensions; it is detachable, reversible, adapted to any kind of mounting, reworked by an individual, group or social formation (Deleuze and Guattari 1987, 12).

For Guatarri (noted by McCormack 2003, 496), an emphasis on the qualities of affect that are non-representable is crucial in his "refusal to short-circuit the creative potential of the pre-personal force of the unconscious by incorporating it into an oedipal order of signification."[5] For Deleuze (1994), creating this cartography involves a total rejection of the central components of representation, which are *analogy*, *opposition*, *identity*, and *resemblance*.

For what I want to do here, this rejection means refusing either masculinity or fatherhood as an *analogy* for men. To do this requires avoiding positioning fathers in *opposition* to mothers (men in *opposition* to women), it means removing the *identities* of fathers from fixed time and space, and it means not setting up the actual father in *opposition* to the ideal or adequate father. It also requires avoiding suggesting a *resemblance* (or lack thereof) of fathers to heroic or historic notions of men. To do otherwise is to play a role in the management of men and their bodies. By looking at fathers from different angles of encounter rather than modes of representation, the writings of Deleuze and Guattari, Connolly and Gibson-Graham begin to open new options.

The poetics of this come from the emotional geographies at play in the awkward spaces of fathering. And here, a return to Billy and Cosmos' Café provides a poignant illustration of the poetics of becoming other.

The sun is now dipping behind a low hill that holds the western boundary of La Mesa and a chill descends upon the café. Billy and I are sitting in a very public place. The tears track silently down his cheeks as he talks about his dad.

"I learned a lot from my dad from a very early age," he laughs at the irony of this, given his earlier story about the shotgun, "he was a tough son-of-a-bitch, but very affectionate too. He was, and this is something that my wife sees in me 'cause I am like my father …"

Billy pauses before continuing with emphasis.

"Because I am like my father. But he worked way too hard. That is one of the things I always remember, he worked so much that we didn't do a whole lot together and, and eh, …"

5 Oedipus as the father killer is a "ready-made tracing" which "always comes back to the same," and deflects us from a cartography "that is entirely oriented toward an experimentation in contact with the real" (Deleuze and Guatarri 1988, 12–13).

Another pause.

"I lost him two years ago later this month, as a matter of fact, and I miss him."

Tears flow freely.

"I miss him a lot."

The conversation returns to Billy's relationship with his older son. Billy's son is eighteen and I join with him in the joy he feels for a life and a relationship, as he understands and feels them, transformed, at least a little. It is an evolving story, a relationship that is *becoming other*. His son is enthusiastically engaged in studies and life at a university in northern California. He'll be returning for Thanksgiving break in a couple of weeks. Billy is excited about this first return home. We talk about his son's 'coming out' in High School and how his struggle with sexuality may have been part of the cutting incident four years prior. As the sun disappears completely and our time together draws to a close, tears well in Billy's eyes once more as he talks about his evolving relationship with his son:

Now
we enjoy together
we enjoy going to movies

I can talk to him
a little bit better
now

But I mean
he had such an internal struggle
you know?

I always suspected that he was gay
you never know, you never know, you never know, you never know
and when he came out
it was like
"okay son that is … that's good"

It has always been a struggle for me.
not a bad one
I love him to death.
I've never stopped loving him
and I think I've accepted it

but

just in that statement
in itself
there is always still that, that …

I don't feel totally comfortable with it

My son really surprised me
the last time we …
he says our relationship's changed
a little bit

And with these words, I understand – for Billy and for me – the importance of recovering fatherhood as a continual project. It is my hope that the pages that follow broaden the scope of these poetics, the awkward spaces they form and the awkward spaces from which are they derived.

Chapter 1
Partially Remembered Stories

This chapter is about methodologies. Empirically, the bulk of the book comprises stories of everyday fathers, men whose lives I have followed off and on for over a decade. I try to give meaning to those stories by elaborating – theoretically and empirically – fathering as an emotional work. The book is also about the stories I am told (and that I tell myself) about the institution of fatherhood through newspapers, movies, academic research, and in the coffee shop. These larger stories reside *beside* stories of everyday fathering in the sense that Eve Sedgwick (2003, 8) means when she talks of a "spatial positioning" that is *not* about understanding what is *beneath* or *beyond*. The larger stories – the institutional discourses, the coffee shop discussions and the imaginaries of filmmakers – are important in the ways I co-construct fathering identities. They resonate with the emotional practices of everyday fathering in complex ways. To elaborate these complexities, I employ a variety of experimental textual and visual methods to render a quirky and elusive image of fatherhood and to give fathers a voice. I'll get to the methodological importance of the visual narratives in a later chapter. In what follows, I elaborate the ways I engage the collective experiences of fathers and the emotional work of fathering through ethnopoetry, dialogue and differentiated stories.

Ethnopoetry and Everyday Emotions

> [T]he poetic imagination … here the cultural past does not count … One must be receptive to the image at the moment it appears: if there is a philosophy of poetry, it must appear and re-appear through a significant verse, in total adherence to an isolated image. The poetic image is a sudden salience on the surface of the psyche, the lesser psychological causes of which have not been sufficiently investigated. Nor can anything general and coordinated serve as a basis for a philosophy of poetry (Bachelard, 1964, xi).

The year I was born, Gaston Bachelard published a landmark philosophical piece entitled *La poétique de l'espace* (1958). I think I am attracted more to the title of this work than to its content. I am troubled by Bachelard's allegiance to 'pure' phenomenology and his insistence that culture is redundant in understanding the poetics of the world. I also smart at his portrayal as the "father of the French new critics" and a "debonair patriarch" (Gilson 1964, viii). My putative judgments aside, *The Poetics of Space* is an imaginative rendering of images that escape psychology or rationalism. Bachelard's poetics are focused on spaces of intimacy and immensity (from rooms and closets to shells and the universe) in an attempt to seek the onset of an image in consciousness. Such a project requires moving beside time while recognizing the potency of material space and time in the rendering of experiences; it is a moment where time ceases to imprint memory, and space and emotion are everything.

Methodologically, it is my hope to engage in the pages that follow men's stories and voices at a moment where space and emotion are everything. Part of this comes through technique. For example, I use representational techniques that are different from traditional social science that relies on theoretical exorcisms peppered with grey columns of interviewees' transcribed quotes. The previous chapter began with something different from a traditional epigram; I began with a poem created out of Billy's discussion of the most traumatic moment in his fathering. If, as I believe, poetry can be an emotive construction of language, then my arrogance is to re-visualize, contort and arrange Billy's words to create something that speaks to his emotions. I call this ethnopoetry because it is derived from the transcription of an interview that is part of a larger ethnography. Creating poetry out of interview transcription is not new, although it is hardly an accepted form of social science practice. Riessman (1993), for example, argues for using different textual strategies to elaborate stories and narratives. Joanna Roberts (2005, cited in Jackson and Russell, forthcoming) transforms interview transcriptions into poetic stanzas to emphasize repetitions and the rhythmic isolation of words. Roberts argues that this form of poetic representation is strongly guided by the interviewee in the sense that the line breaks and other poetic features of the text closely follow the qualities of the interviewee's oral account.[1] Jackson and Russell (forthcoming) point out that although this is an innovative attempt to push the boundaries of textual representation, some researchers may argue that it reads too much into the text and that the balance of power slips too far in favour of the interviewee. As a potential problem, this bothers me less than whether or not the ethnopoetry conveys affect(ions) that broach a larger series of conversations.

My use of poetic stanzas is to provide a parsimonious rendering of the emotions that exceed the text. I work at the words, pushing them to reveal the emotional power of a conversation. I want to get at the embodied power that resides within the words. It is about the language of looks, twitches, grimaces, tears, laughter. It

1 Roberts (2005) includes audio material in a CD-ROM to facilitate comparison with her poetics. To do so here would contravene the ethical guidelines of the Human Subjects Review Board at San Diego State University. Roberts' project remains unpublished and so, in all likelihood, does not contravene ethical issues that surround research in the UK. Any published research involving human subjects at US universities requires oversight by an Institutional Review Board (IRB). Prior to embarking on the *Awkward Spaces* project I wrote out an extensive protocol regarding how I would conduct the research; this document included possible benefits or harm to participants in the project. In order to protect the anonymity of participants as far as possible a number of measures were suggested, including the promise not to disseminate audio files.

is about connections and places and past remembered conversations. Michel de Certeau and his colleges (1998, 253) put it this way:

> Priority goes to the illocutory, to that which involves neither words nor phrases, but the identity of speakers, the circumstance, the context, the 'sonorous materiality' of exchanged speech.

Irene McKinney, West Virginia's Poet Laureate, is adamant that poetry should be conversational and parochial, "written out of a grounded place-anchored life" (McKinney 1989). Ethnopoetry is a halting and partial – and yet sincere and promising – attempt to represent the non-representable of conversation.

As a narrative experiment, I use different genres – mixing poetry with dialogue and academic discussion – to engage in a more forceful way with the men whose work as fathers I genuinely want to celebrate. Through the silences that join and link the narrative and the poetry, it is my hope that emotions foment and reveal themselves. These are the collective experiences of fathers elaborated through remembered stories, the content of which is always partial and incomplete and always generative and creative. What is true and incontrovertible about these stories are the emotions that render, and are rendered by, memories.

Embodied Collective Memory

Ultimately it is from the work of fathering that an idea of fatherhood emerges. To engage this collective experience I ask fathers to tell me about their work as fathers, as they remember it. I do not look for accurate accounts but for emotive, embodied responses that are mediated through memory and the telling of stories. Part of my assumption is that collective experiences are palpable and important in the co-creation of the subject-self-father and the framing of fatherhood as an institution, what Judith Butler (1997) calls "subjection." By listening for *expressions* of transformation – fathers' *becoming other* – as a series of embodied, emotive and collective experiences, I am concerned about "emotions that exceed the fund of subjectivities" that are provided by larger institutional frames.[2] These expressions

2 From Deleuze (1990), Massumi (1992, 12–13) argues that through the contortions of the ideas that are available through memory comes *expressionism*. Similarly, Gibson-Graham (2006, 51) seek "expressions of fugitive energies" that move beyond institutional frames. By so doing they find, for example, expressions of "care for the other, concerns for justice and equity … and calls for new practices of community." These expressions are elaborated from collective experiences and embodied memory. In the sense elaborated well by Gilles Deleuze and Felix Guattari (1987) the framing or striation is smoothed yet again and then striated once more to the extent that the process of *expression* becoming *content* and vice versa and is, in and of itself, simultaneously an illness and remedy, a remedy and an illness. Content (the institutional frame) is what is overpowered; expression is what

come from memory that is understood neither as "a single narrative or community consensus, but as the shared enactment and reenactment of experiences and feelings in place" (Gibson-Graham 2006, 211). For Deleuze (1990, 311), a study of Spinoza's expressionism in philosophy leads to a sense that "imagination corresponds to the actual imprint of some body in our own, and memory to the succession of imprints in time. Memory and imagination," he suggests, "are true parts of the soul." According to Henri Bergson (1988), stories emerge through entanglements of affect, emotion and memory. And, importantly, memory is immediate (virtual). Bergson suggests that memory is not about the past, but is a continually embodied process working towards a future (1988, 65). As Deleuze (1989, 207) notes:

> memory … is the membrane which, in the most varied ways, makes sheets of the past and layers of reality correspond, the first emanating from an inside which is always already there, the second arriving from an outside always to come, the two gnawing at the present which is now only their encounter.

Fathering stories begin at their middle-point, they spread out, seeking direction from bodies, home places and youthful origins, and they are complicated in a vast arena of experiences and emotions: a compass rose of memory.

An example may help clarify the complex relations between emotions and memories that are neither a single narrative nor a consensus, what Addelson (2005) calls "embodied collective memory" (cited in Gibson-Graham 2006, 11).

Meet Ed. I've known Ed for about thirteen years. At sixty, he is ten years my senior and that time carries with it a wisdom that outweighs his relatively short time as a father. His son is eight at the time of the conversation I relate here. It seems to me that the pendulous weight of Ed's years is held in physical abeyance by a strong, six and a half foot stature. Ed's face is squared by a full head of grey hair that is cut short so that it bristles on top while flopping slightly over his brow. He and I are enjoying supper at a local restaurant when he offers me an example of embodied collective memory. Ed's fathering story shows up in Chapter 2, but for the moment he is talking about the relations between memory and emotions.

"I remember thinking the other day," Ed says, "that I have too many examples of why not to trust my memories … Like I can remember a traumatic event when I was four, maybe three, and, and it was around … I don't know what it was … I think my sister – see, I can't even remember if it was my sister – maybe my brother." He laughs. "We were all chasing each other around," he pauses and laughs again. "And then there was some trauma about a coat-hanger that had come off of a door knob and we were chasing and stuff and it got caught in my eye. Okay, and I just remember a bunch of weirdness around that." He looks at me pointedly. "I don't remember specifics, just that general intensity of emotional trauma. And I remember bringing it up to my sister five or six years ago and, ah

---

overpowers (Massumi 1992, 152). Deleuze and Guattari (1987, 41–44) sometimes refer to this as a double articulation (see footnote 4).

…" Ed pauses for emphasis. "The only thing she remembered was that it was in her eye not mine."

Ed pauses again, and smiles at this encounter with his sister and her memories. Expressions of care and love are often elaborated through shock and surprise. Embodied collective memories of this kind create spaces for relations that comprise everyday life and emotions.

Ed continues:

> but it doesn't matter
> it makes no difference
> so,
> what I don't trust are the actual facts around what happened
> but the emotional intensity around them is the truth
> and that is what affected me
> that is what affects me now

If the emotional spaces of fathering are an important aspect of transforming men's lives, then I get at them, at least in part, through memories. To do so, I listen to fathers with an ear towards repetitions, rhythms, echoes and cycles.

I've been listening to fathers talk emotively about their relations with their children and their families, including their own fathers, for many years now. Sometimes, I get involved in a protracted discussion. Mostly, I listen. When I feel that there is a degree of trust between us, I ask for a fathering history: an oral history with a focus on the man's fathering, whomever modelled fathering for him, and his thoughts on fatherhood.

Mary Chamberlain and Selma Leydesdorff (2004, 227) point out that oral histories are always about engaging memory through lived stories. Memories and narratives, they suggest, are raw ingredients that yield insight into individual emotional adjustments to cultural transformations. Memories are personal and emotionally charged, but they are also social and, as collective constructions, they contrive larger imaginaries of family and community. Ultimately this is what I want to get at with this book: men's lives placed beside their fathering, beside their families, beside their communities.

As I hope for clarity in the pages that follow, I am suspicious of the notion of embeddedness – fathers embedded in families, embedded in communities – because it implies a hierarchy of scale that is disempowering to some and empowering to others. I live beside rather than under, or through, or between other people. Other people's memories become incorporated into my own – sometimes, like Ed's coat-hanger, they become my own – and this is especially true of family members. And so, here, the fuller aspects of my fathering project unfold. This book's stories about the remembered spaces of fathering – even from a larger sample of several hundred stories – do not suggest for a moment a common history, a community consensus or a coherent theory. That is not the intent: there is no social generalizability from stories of these kind but there are, through the intensity of the emotional recollection, eidetic pointers to larger changes and transformations. As Ed suggests, emotional intensity is a truth that affects.

As part of memories, stories are frameworks that help me to make sense of my world as a father, as a man, and as a human. Stories are always mediated and censored, publicly and privately, officially through institutions and unofficially through personal recall. Stories permit and fashion recollection that is socially and culturally specific. As with Ed's recollection of the incident with the coat-hanger, memory raises issues of validity and how things are represented, but what is non-representable – the emotions associated with remembered event – is incorruptible because it is always different.

In emphasizing the collective embodiedness of memory, Chamberlain and Leydesdorff (2004, 229) point out that "[t]he language, images, contours and colours of memory, the guidelines and judgments, dreams and nightmares, omissions and commissions that inform the imaginative act of remembrance are … shared and therefore social." The symbolic structures of culture are embedded in language and, as a consequence, narration. The metaphors and similes that people use – their rhetoric, their stereotypes and their sayings – all suggest values, priorities and ways of knowing and interpreting the world. These tools for telling individual stories are also frameworks for memory and as frameworks they reflect, refract and represent the narratives through which a society, a place, a nation tells a story about itself.[3] Communities and families, then, become sites of belonging, an embodied imaginary gleaned from oral histories. The oral histories in the pages that follow focus on the emotional work of fathering and the spaces that connect that work to families and communities.

Spatial Framing

At some level, the work of the father must have an image in order to act. This may be conceived as a structure – a spatial frame – from which actions emanate. Fathers are almost always caught in some form of spatial framing. And the institution of fatherhood is also embroiled with framings of this kind. But the actions and assemblages of fathers are also formless, what Deleuze and Guattari (1987) call "*bodies without organs*". In other words, fatherhood is an institutionalized network of virtual relations and, as such, it is difficult to conceive of, and to push against. Nonetheless, this institution is immediately actualized as the material work of fathering wherever one of fatherhood's working images – the organs: the patriarch, the loving dad – go. These working representations convey action, they bring to each place and time – to each Deleuzian spatio-temporal coordinate – a relation

3 Chamberlain and Leydesdorff (2004, 230) point out that "[w]hile the personal narrative may be seen as the property of the individual – intrinsic to and defining of the individual – the plot that it follows and the themes that are woven through it may reflect and conform to the cultural narratives to which any one individual is exposed at any one time."

that fundamentally changes those bodies' social and physical realities. The relation *is* father as an immanent social agent.

Meet John. John's story helps me understand spatial framing as a methodological contingency. It is a story that, from one viewpoint, ties down a seeming precise and coherent image of fatherhood but, as I hope to show, it is also actualized in ways that belie coherency. And the key to unpacking this frame lies in John's emotions and his emotional work as a father.

I've known John for about five years and it seems that we've been talking about our fathering practices since the day we met. Today, I have my voice-recorder with me, and we're engaged in his fathering history. John is talking about when he met his current wife.

"I wanted to find someone who wanted the same things that I wanted," he says, "family, the hearth, the home, the security. I didn't really know what it was. I've never experienced it, but I knew that I wanted it."

John and I are having lunch at a small café atop some high cliffs above Torrey Pines beach in La Jolla over-looking the Pacific. This point on the cliffs is always windy and for years has been the launching point for hang-gliding enthusiasts. I wonder, with some amazement, at the ways men – like John – throw themselves into relationships – like fathering – with little sense of security beyond the stories, the myths, the fantasies of what a family might be like. And, like the hang-gliders that are tipping off the cliff in front of us, John's sense of family and community teeters on the brink of destruction and then soars in an upwelling of emotion, responsibility and, for John, security.

"I just feel so much more secure having a family, having a wife, having a house," says John. "It is not the things – it is not the house it is not the wife, well it is the wife 'cause she is a wonderful, wonderful woman – but it is not the things like the house and the car exactly, in and of themselves. It is what they mean. And what they mean to me is *place*." John emphasizes the last word. He knows I am a geographer.

"You know, this is my place in the world. And," he laughs, "it is my cave."

Trained as a journalist, John now works at a local university. He is well educated. It might be argued that he embraces, at least in part, the Heideggerian notion of being-at-home-in-the-world and dwelling-as-enclosure: "dwelling as a limitation of the event in space through the insistence on holistic closure, autarchy, and self-sufficiency" (Harrison 2007, 642). But he also embraces Derrida's more open formulation that aligns with the idea that "the hearth [*le chez soi*] of a home, a culture, a society presupposes a hospitable opening." This opening is not necessarily exclusive to another like myself but, according to Derrida, also to "an other who is beyond any 'its other'" (Derrida 2002, 134 and 364, cited in Harrison 2007, 642). For John, this openness finds form in strong emotional attachments.

"This is my cave," John goes on,

> I have ownership of this cave
> and the things in this cave are mine

and the people in this cave support and love me
and I love and support them
that mutual love and support is extremely powerful
it may be the most powerful thing on the planet
what is more powerful than the family? the clan?
this is the basis
to me
this is the basis of what humans do

By listening for *expressions* and emotions that exceed the fund of subjectivities afforded by metaphors such as 'the cave' I identify with John's calls for care, for justice and for equity within families and communities, and his calls (perhaps) for new fathering practices as they move towards potentialities that have arisen out of a particular subjection (Gibson-Graham 2006, 51).

Our sandwiches arrive, much to the excitement of Maggie, who accompanies me on many of these interviews. I tie her off well out of reach of the food. John and I talk some more about the idea of family and the security that it provides.

"… that helps me in my parenting, I mean, it helps me to feel so grounded. I can feel so grounded in a place that the place becomes part of the fathering, the place becomes part of the parenting. Because it is this secure place."

"Is it about security primarily? In a broad sense?" I ask.

"It is. In one way it is about me feeling secure and the more secure I feel the better a father I can be. But there is another angle, another part of it too, and I am at a little bit of a loss about it but it is, somehow, that this being the family place – the hearth and home – that I am certain of, that I am sure of, makes fathering my children there so much easier. It is a lot about security but it is security for my whole family. It is really like …"

John takes a long pause and into the silence I interject: "Is it emotional security?"

"It is, yes, absolutely," John replies. His placid features scrunch into a self-conscious smile. "Emotional security. That is really the only kind there is, frankly."

"It is the ultimate family fantasy," I suggest. We both laugh.

I ask John how his notions of family, hearth and home are connected to community.

"I've always been kind of contemptuous of community," he says, "I don't know why. Mainly because I never really felt like I fit in," he pauses.

"My wife is very good [at community]. She's always lived in La Mesa. She knows everybody there. She is really secure there. My kids go to the same grade school that she went to. She has trod these paths and streets of La Mesa forever. Both literally and figuratively it is very powerful for her. It is a very strong thing for her. And I have gradually come to see how our place, literally *our place*, in this community that we live in – this part of La Mesa, this school district, this neighbourhood – provides a strength and security."

John smiles wistfully.

"Again, back to the security. How much strength and security there is in that!" he exclaims. "You know, I am kind of a stand-offish person. I am an introvert, left to myself, because of my own fears and insecurities and so forth. So it takes me a while to feel part of a community. You know, as my wife just kind of embraces the community, I stay aloof. However, even though I am aloof I am still there and my presence has been growing and, against my worst wishes, I have become part of the community anyway. And I see that it is a very good thing. It feels very good. Even although I am cynical about it … But I really see the benefits to me of this thing, this amorphous thing I am trying to talk about in the world: my family and my parenting and my, you know, … I have a place."

He smiles generously.

"Community is just like an extension of that. I actually work hard at this stuff. I'll tell you what, I have to work to be a good parent, I have to work hard to be a good husband, I have to work hard to be a part of the community because I am … I am … afraid of it. And I really want to do something else. I want to do something different. But that is not what is good for me."

I am intrigued by John's fears, and why he thinks that community is good for him, and when I ask he elaborates by talking through his notion of fatherhood in today's society. To a large degree, his comments mirror well Connell's (1995) 'crisis tendencies', as elaborated in the last chapter, to the extent that they raise the notion of men/fathers on the brink of hysteria, looking into a pit from which they can derive very little respect or self-worth.

"I see fathers, in particularly, in our culture, as way too often being characterized as superfluous. And I really don't like that. I think it has got a little bit better because there seems to be a lot more focus on parenting and children and so forth, at least in a superficial sense … I've seen this thing, you know, in mass media fathers are seen constantly to be characterized as clowns. They, the media, want to characterize the father as goofy. Now, it did not used to be that way. In the 50s and early 60s the father was characterized as strength and wisdom. And then in the 70s and 80s everything got weird … And you started having these sitcoms where the father was the goofball. And that became almost ubiquitous. I mean *The Cosby Show* and some of the other things. I see them in cartoons – the cartoons my kids watch – you know, the father is often the goofball. [Me and the kids] can joke about it a little bit … I think things have got a little bit better. There is a little bit more respect for fathers and yet, you know, [I worry] that the father's role is [still] a superfluous thing."

"Where do you think it should be?" I ask. John draws in his breath and looks quizzical. I try to clarify my question. "If you had a perfect society, what would it look like for fathers?"

John exhales slowly. "In a perfect society, every child would have a father … every child would have a father active and critical in his or her life. And in a perfect society, the father would be both required by law and by morality and conscience to be actively involved in a positive way in his children's lives… And that can look like a lot of different things, I mean, you know, I don't … we have a lot of friends and all the parents do it differently. You are a different father than I am. We have friends who parent their children differently than I do. And you know I tell my kids when they are at my house or other houses, you know, I say 'Listen, each household family is different and you've got to have respect for other ways.' What I cannot respect is when the father does not take a role … So, so my perfect world really is that every child has a father and every father be actively and positively involved in the child's life and, you know, however that might be. That could look differently: I don't want people to adopt my sense of morality or religion or parenting skills or whatever. I don't want them to do anything harmful but there are so many different positive ways that people can interact with their kids."

John picks up his sandwich. This seems like a good point to finish our discussion for now. There are numerous whitecaps on the ocean. The wind is picking up and I am worried about the audio quality on my recorder.

"Well, that is it for now, John," I say, "unless there is anything more you want to add. I appreciate your time …" I finish off my sandwich.

John wipes the corners of his mouth and places the napkin on his empty plate, under the cutlery so that it does not blow away.

"Now tell me something," says John, "what is this book you want to write?"

I tell John about the *Family Fantasies* project and the questions that it raised for me about the so-called domestication of dads and how that relates to the ways fathers' nurture.

"You touch on a really important point," John is once again animated, "is the recorder still on? This is really important to me. You know, there has always been this idea that men cannot be as nurturing as women. This is not true. The fact is that men nurture their children in different ways. A lot of the things that we hear about parenting are standards that are defined by women. Now, men and women look at things in different ways. One is not bad and one is not good. They are different. Women look at things in different ways. This whole nurturing … the idea of nurturing has taken on a female definition of late. There is female nurturing and there is male nurturing. There is mother nurturing and there is father nurturing. They are just different. And the father nurturing looks … for me it has a lot more to do with safety, security, and example. It is more action oriented." John stops for a moment and contemplates what he has just said. He reconsiders, "okay, it is just different."

John articulates an opinion that is expressed, perhaps in a more muted form, elsewhere. Common wisdom, bolstered by academic insight, often defines a father's relationship and involvement with his children as a form of co-parenting that is interdependent with, in opposition to, and at times less than mothering. The media and social science literature has trouble defining this difference. John is correct to the degree that fathering as a practice and fatherhood as an institution are difficult to define without recourse to mothering and motherhood. The Western media are replete with stories of caring, nurturing, domesticated dads who are fighting the strictures of the workplace to spend the same amount of time with their children as mothers (e.g. Sauer 1993; Pryor 1999; Crary 2008). In addition, a burgeoning social policy literature is focusing on men's co-parenting skills (Lamb

1997; Brandth and Kvande 1998). The point I want to make – and I think John does too – is not that there is a convergence towards a co-parenting model of the family, but rather that media coverage and social science research compares levels of fathers' family involvement with mothers' involvement because mothers are the benchmark for norms in fathering (Doherty et al. 1998, 278). This is not co-parenting, but mother-defined parenting. I am not suggesting that fathering can be defined in isolation from mothering, but I am concerned that the idea of the father is constituted in parallel or in opposition to the idea of the mother and, as such, does not account for the imprecise and hesitant day-to-day work of fathering as resistance and negotiation. These are positions that tend to avoid issues of power and dominance, and often soft-pedal the nuanced and differentiated emotional bonds between fathers and their children. Rather, it seems reasonable to argue that mothers and fathers 'double-articulate' the work of parenting. Deleuze and Guattari (1987, 41–44) use the term double articulation to imply that there are diverse ordering or defining processes at work simultaneously in the same being, but at different levels (Due 2007, 131).[4]

As part of the exploration of fathering as a double articulation it is important not to miss the doings of men, women and children as daily emotional practices that are negotiated, contested, reworked and resisted differently in different spaces. Of central importance is the assumption that these spaces are highlighted and questioned when a father's position in a family (his roles, responsibilities and power) is put in jeopardy. Many fathers admit – like John – that, at the very least, their normative notions of what it is to be a father are continuously challenged through the course of their fathering. With this project I seek to uncover some of these differentiated stories.

Differentiated Spatial Stories of Fathering

Bound as a set of everyday stories of fathers' lives, the book is about a continuous unbinding – in terms of, say, the de-territorialization of traditional patriarchal ways of knowing – and a rebinding – in terms of, amongst other things, the re-

4 According to Due (2007), these levels relate to different types of organization. The first level is a dynamic interaction of simple elements that Deleuze and Guattari (1987, 41–44) call *molecular* organization. The second level is an ordering that relates to categories, which they call *molar* organization. The composition of these two ordering principles forms a *stratification*, which can be undermined (smoothed) by *deterritorializing* forces. These forces constitute a *line of flight* that affects the whole 'assemblage' of stratifications. To account for events or processes by invoking structures such as class, mother, father, family or culture is, according to Due (2007, 132), "to overlook the double articulation of any stratified being as both molecular and molar, thereby missing the way that any event or process is involved in several levels of reality at once." For Deleuze and Guatarri (1987, 44), double articulation sometimes coincides with content and expression (see footnote 2), because content and expression are often divided along molar and molecular lines.

territorialization of the work of fathering as an emotional practice. It is about recovering fatherhood through emotional geographies – through, if you will, tender mappings – that eschew traditional ways of knowing *the* father as a monolithic construct contrived in patriarchy, and other past (and continuing) sensibilities.

Doreen Massey (2005, 25) points out with Michel de Certeau that there is an important connection between the modern scientific method that emerged with writing (letters and journals) from 19th century empiricism and the creation of a blank space (de Certeau's *espace propre)* not only for the objects of knowledge but also for the act of writing and representing. For de Certeau (1984), the importance of spatial stories, with all the nuances of narratives through movement and surprise, is lost to science as "the writing of the world" (see also Curry, 1996) and to map-making as a practice that attempts to control space. These, he suggests, are detached, scopic representations of people and places, which are very different from walking in a city block or taking a tour, spatial practices upon which de Certeau places a high value. In an aesthetic that rehearses later arguments by Elizabeth Wilson (1991) and Iris Marion Young (1990) for the celebration of difference, de Certeau particularly likes the surprise that may accompany the *flanêuse* as she walks through city spaces. But de Certeau's insights falter on the cusp of representation and the strictures imposed through a different kind of framing. In order to disrupt this spatial coherency and rekindle the dynamism and differentiation of life, Massey (2005, 26) argues that we need to get beyond de Certeau's (and Laclau's) "equation of representation and spatialization" because the spatial is not stable; it cannot flatten the life out of movement and time. What is required for the opening up of the political, argues Massey (2005, 25), is simultaneously a space of freedom and memory (from Bergson, 1910), dislocation (from Laclau, 1990) and surprise (from de Certeau, 1988). The spaces of my encounters with fathers require similar characteristics to derive a respect that removes them from the vestiges of containment, generalization and order in social science research, from the oppressions of privilege and entitlement that emanate from the academy, and from the circumscription of journalistic denouement. I am trying to create a differentiated field of opening and experimentation, of playfulness and feeling.

PART I
Framing

Chapter 2
Fathering Frames: Some Histories and Geographies

> He was shut out from all family affairs. No one told him anything. The children, alone with the mother, told her all about the day's happenings, everything. Nothing had really taken place in them until it was told to their mother. But as soon as the father came in, everything stopped. He was the scotch in the smooth, happy machinery of the home. And he was always aware of this fall of silence on his entry, the shutting off of life, the unwelcome (D.H. Lawrence, 1976/1913, 62).

Something happened to the spaces of fathering as the industrial era progressed. D.H. Lawrence's rendering of the marginalized father figure, distanced from his family and shut out of its life, continues as an increasingly poignant representation through the 20th century and is augmented – to some extent demonized – by feminist attacks on patriarchy and the rising spectre of the dead-beat dad in the 1980s and 1990s. The story of fathers increasingly distanced from family life is so well worn in *fin de siècle* academic (Blakenhorn 1995; Popenoe 1996) and popular literature (Bly 1990; Faludi 1999) that I do not intend to dwell on it here. It is a tired story; as tired as the fathers who, from the early part of the century, are represented variously as drunk and violent (Lawrence), as lost souls looking for mythic redemption (Bly) or as overwhelmed and stifled by larger political and economic changes (Steinbeck, Faludi). Rather I want to put the stories of marginalization and distancing together with other stories as a series of spatial frames so as to complicate, just a little, the histories and geographies of fathering. My intent is to complicate the spaces of families by celebrating the changing contexts of fathering geographies along with the changing contexts of communal households and community spaces.

The book's introduction and Chapter 1 open stories of fathering, conceptually and methodologically, wherein the space – of memories, of communities, of hearths – is elaborated as an important part of narrative. With this chapter I discuss the contemporary Anglo-American story of fathering as writ large in common wisdom and through academic literature (with a brief and quirky nod to popular literature). To some degree, what I provide is a familiar set of stories with a trajectory that is well known. The origins of these stories are not so much lost as impossible to construct.[1] Stories without beginnings or endings nonetheless start somewhere;

---

1 The writings of 17th century philosopher Baruch de Spinoza (1996) as read through 20th century philosopher Gilles Deleuze (1988, 2005) suggest that nothing (e.g. a story)

my choice is to start with Enlightenment thinking because of the ways I believe it carved – with unsteady hand – the contemporary stories of fathering through and out of patriarchy. Perhaps the Enlightenment is simultaneously the beginning and the end of patriarchy and so I choose to begin my tale of the emotional work of fathering from this middle-point.[2] Given that I can no longer assume a coherent all-knowable reality, neither can I depend on immutable forms, meta-narratives, transcendental meanings or material representations. If fathering, as a work and an affect, is not dependent on origins and beginnings, it nonetheless calls into question the works of patriarchy. And, in that questioning, there is a responsibility for forms of oppression and domination that accrue to individual fathers. I hope to show in the following chapters the ways that patriarchal origins are replaced with, and placed beside, other assemblages of content and expression (Deleuze and Guattari 1987), but before I get to that place I want to create in this chapter an image, a partial tracing of the content of fatherhood. And so, as the chapter progresses I move from the mid-18th century through the 19th and 20th centuries stopping at a few places to describe briefly the landscapes of fathering as I see them. It is not a complete story and it is not an attempt to wring out some spatial axiom from which fatherhood finds form, even though sometimes it may feel that way. Rather, it is my meanderings through a dense and circulatory literature from which no answers arise, from which there is no beginning or end, and from which some musings take form before evaporating again, and from which some surprises emerge.[3]

---

belongs to anything else (e.g. a meta-narrative), and that all stories start somewhere in the middle.

2 In what follows I want to argue that Enlightenment thinking (and the doings of Enlightenment thinkers) were a deterritorialization of parts of patriarchy and a reterritorialization of other parts.

3 The chapter is also my attempt to critically circumvent a form of social science teleology that has emerged over the last half century, particularly from within the discipline of sociology. That emergence, I argue, is part of a larger story with a beginning in modernity and a well-defined and potentially problematic ending. Let me footnote this teleology so that I am quite clear about the literary and theoretical contexts that I choose to elaborate in this chapter. Beginning in the 1980s and continuing today, sociological work for the most part focuses on co-connections between fathers and mothers in light of the changing economic context of women (cf. Coltrane 1996; Lupton and Barclay 1997; Dienhart 1998). Based on large survey data and, on occasion, case studies, this work seeks to reconstruct the context of fathers in contemporary Anglo-American society. As important as this work is in terms of the creation of particular kinds of empirical data and the generation of some interest by policy-makers, it nonetheless suggests a certain teleological orientation towards the increased incorporation of men into families in terms of domestic responsibilities that move from the distant breadwinner, to the helping husband, to today's father taking responsibility. Even sociological studies that explicitly pursue post-structural perspectives (e.g. Lupton and Barclay 1997) seem nonetheless to embrace a linear trajectory of fathers heading towards some domestic idyll. This is certainly an important story of partnerships within

I begin with two brief fathering stories, one from the beginning of the Enlightenment era in the United States and one from just after in the United Kingdom. They involve famous men whose private family lives are footnotes at best. I then take some time with the ways fatherhood is shaped and constructed from the beginning of the Enlightenment and why these ideas foment into something mythic.

I explore four interrelated trajectories with this chapter that follow through the balance of the book. First, I want to highlight some prominent ideas of fatherhood. Second, I suggest that the emotional work of fathering offers a different encounter with fatherhood. Third, I want to move knowledge of fathering beyond its definition in opposition to mothering. Fourth, I begin to move – in a stalling and imprecise way – the body of knowledge on masculinities with a focus on fathers that specifically addresses how the practice of fathering and its emotional work is learnt in relation to women and children, and shaped by produced and occupied family and community space.

What I begin with in this chapter is a journey that begins with the suggestion that early modern writing about the work of fathering was less about reason, power and property, and that although certain problematic spatial frames arose from this writing, the real enlightenment and progress emanated from emotions.

Forgotten Histories and Geographies of Fathering

Thomas Paine is an erstwhile 'founding father' of two modern nations. He is best known for lengthy, popular, evocative pamphlets that kindled, sparked and provided fuel for the flames of revolution in 18th century America and France. Paine

---

communal households but it elides the political in the sense that fathers are constructed within a particular paradigm that forecloses upon the potential for new and transformative fathering identities. Based primarily on surveys, census data and other documents, this work speaks to aggregate changes and makes substantial inferences on lifestyle changes. But surveys and other records are inevitably bloodless. They tell me a great deal about the business of mothers' and fathers' lives, but almost nothing about the emotions of it.

In denigrating surveys, I am not talking about the very important local geographies of mothering (Dyck 1990; England 1996; Holloway 1998; Aitken 2000d) or the larger stories of reproduction, which until recently remained hidden to academic scrutiny (Mitchell et al. 2004) or to the importance of gender equality within communal households (Fraad et al. 1994; Gibson-Graham 1996). The stories that I want to elaborate upon are hugely related to these other stories, in the sense that they reside beside them and are co-created through them. The larger social science story that is problematic is, for the most part, empirically driven and culminates in a convergence between mothers and fathers in terms of responsibilities and child-care (Crary 2008). This is not a new teleology. Pleck and Pleck (1997) argue that the father has evolved from the distant breadwinner of last century, through the genial dad and sex role model of most of this century to today's father as equal co-parent. Unlike the stories of day-to-day mothering and the importance of reproduction, the conclusions of this story are already written: in a variety of forms, fathers become domesticated.

is not known as a father of America in the same sense as George Washington, John Adams, Thomas Jefferson and others who later became presidents of the nascent world power. His birth in Britain notwithstanding, Paine had no aspirations for high political office. His consummate focus on Enlightenment ideals created many political adversaries over his life, and his lowly working class origins and bouts of depression and alcoholism did not ingratiate Paine with other leaders of the American and French revolutions. Nonetheless he was a great writer for, and champion of, the so-called common people. His *Common Sense* (1776/1979) enflamed American sentiments against George III and his *The American Crisis* series elaborated "the times that try men's souls" (Paine 1777/1979, 55), while pushing for an enlightened democracy. What is less known about the childless Paine were his propensities towards fatherhood.

Biographers sometimes speak to Paine's quirky relationship with women and their families. For example, in an expansive rendering of Paine's life, Craig Nelson (2006, 43) briefly notes a connection with a family in Lewes, a small town in the southeast of England, where as a young man he worked in Excise. Paine was very close to Samuel and Esther Ollive and at the age of 34 moved in with the couple and four (seemingly unrelated) children who lived above their store. When Samuel died in July 1769, Paine tried to help Esther with the business. In March 1771, he married Samuel's only daughter, the24-year-old Elizabeth, but the marriage was never consummated. Paine never said why, noting to his lifelong friend, Clio Rickman, that "It was nobody's business but my own" (Nelson 2006, 43). His marriage to Elizabeth was his second; the first ended in the death of his wife and child in childbirth. Of this event he wrote at 23 that "there is neither manhood nor policy in grief" (Nelson 2006, 38). These are, I aver, telling remarks of a man at the cusp of writing the birth to two modern nations, of reworking notions of civic responsibility and of conceiving a democratic notion that would grow global over the next century and a half. They are telling in that they eschew emotions and consign fathering and family to a private space that is nobody else's business. They are, perhaps, the seed of the story of modern fatherhood.

With the dissolution of his marriage to Elizabeth, Paine devoted himself to his work to the extent that he may well have chosen celibacy. If this is so, it is perhaps another telling context of the ways manhood was constructed from this time onward, with work divorced from, taking precedent over, or at least spatially removed from sex and children. Even although biographers enjoy commenting on Paine's supposed celibacy, that story does not detract from his fondness of family and children. While in Paris writing his famous *The Rights of Man* (1783/1970), Paine stayed at the house of his publisher, Nicolas de Bonneville. Paine lived for five years with Bonneville, his wife Marguerite and their three sons. The time was part of his recovery from illnesses that were in large part a product of his incarceration during the French 'reign of terror'. A connection was forged with the Bonneville family that continued to the end of Paine's life.

On returning to America for his retirement in 1802, the 65-year-old Paine first set himself up in rooms in New York and then, with a yellow fever epidemic in full swing, he left for his farm in the French Protestant refuge of New Rochelle, a day's ride to the north.[4] Paine's intimate connection to the Bonneville family is made clear in August of 1803 when Marguerite and her three sons left France to join Paine in America.

There is an important geography to the last few years of Paine's life, although most commentators focus on his continued political wrangles, his heated disputes with publishers, his sinking into poverty, his madness and drunkenness, and his growing infirmity. At first he moved back to New York to be with his adopted French family and to find work for Marguerite as a French tutor. When this failed, he moved the family to New Rochelle while, at the same time, taking steps to make sure the boys were properly schooled. One of his primary concerns in moving back to New Rochelle was that his new family would find an appropriate haven away from the debauchery, grime and squalor of New York City. In time, the cold winters became too much for a physically and mentally infirm Paine and the family once more returned to New York.

In June 1809, at the age of 74, Thomas Paine died. Marguerite and the Bonneville boys were at his side along with some close friends. In his will, Paine leaves a substantial sum of money to Marguerite, one quarter of his New Rochelle estate to Nicolas de Bonneville, another quarter to Clio Rickman, and the final half to the Bonneville boys. Of his last moments, Marguerite writes:

> Seeing his end fast approaching, I asked him, in the presence of a friend, if he felt satisfied with the treatment he had received at our home, upon which he could only exclaim, 'O!, yes!' He added other words, but they were incoherent. It was impossible for me not to exert myself to the utmost in taking care of a person to whom I and the children owed so much. He now appeared to have lost all kind of feeling. He spent the night in tranquillity, and expired in the morning (quoted in Nelson 2006, 323).

Taken at face value, the quote suggests that the last remarks of Thomas Paine, perhaps even his last coherent thoughts, are not of the nations for which he was known as a founding father, but of a more intimate connection with family. It is remarkable – and hugely in line with what I want to say in the balance of this book – that a man whose fame and fortune (and misfortune) are intricately interwoven with the birth of a new era in the Western thinking ends in thoughts of family.

It is important to uncover these hidden stories of fathering because they are remarkable and surprising, and because they challenge conventional thinking. Elsewhere, I recall the story of William Ewart Gladstone who famously served as prime minister of Great Britain a couple of decades after Paine died (Aitken 2005). Gladstone is known for his spirited rivalry with Benjamin Disraeli, his public educational reforms and his championing of home rule for Ireland. His

4 The 200-acre farm – abandoned by loyalist Frederick DaVoe – was given to Paine in 1785 for his services to the American revolution.

private championing of the rehabilitation of prostitutes – something he began before taking public office, and which continued as his passion throughout his life – is also quite well known. Gladstone entered public office in 1832 and served until 1895 with the exception of a one-year absence. It is an important lack of public presence that rarely is commented upon by biographers, or it is handled as an oblique footnote to his more prominent public notoriety. And it is another hidden story of fathering. Gladstone removed himself from public office to spend a protracted period of time sitting at the bed-side of his terminally-ill infant daughter Catherine, who died in 1850.

Thomas Laqueur (1992), in commenting about Gladstone-the-father-in-private, suggests that masculinity, as a symbolic and mythic form of identity, is overtly determined historically and overly-framed spatially as a public endeavour that contrives a coherent and powerful discourse of stoicism and power. Fathering histories and geographies remain incoherent and inexplicit – if not completely hidden – because they are awkward, rarely fitting the spatial frames of public ventures and power that emanated from Thomas Paine's Enlightenment, if not before. It seems that the history and geography of man-as-father is almost exclusively subsumed under a monolithic public geography constructed in large part by a continued and pervasive patriarchy that includes the space of public authority and its transmission over generations. This public geography lauds the notion of a private moral authority (over prostitutes and children alike), but it falls short of addressing the awkward spaces of what Laqueur (1992) calls the 'public-man-in-private'.

As a consequence, and as part of the problematic and inchoate private/public binary, fathers confront an identity predicament that is for the most part hidden, and is always awkward (Aitken 2005, 223). The awkwardness stems from unclear models of how men should be fathers, which is not necessarily a bad thing, as well as a lack of recognition of what constitutes the work (and especially the emotional work) of fathering, which is without doubt a very bad thing.

An Attack on Patriarchy and Praise for Passion

Flicking the pages of 18th century Enlightenment writing suggests an important attack on patriarchy with an opening for change in a number of places. Whereas democracy for all is suggested, there is also the suggestion of familial closure through an insidious spatial framing that contrives the wisdom of the nuclear family. This particular 'writing of the world' by Enlightenment scholars seemingly removes some real life emotional dynamism from the work of fathering, but in what follows I want to suggest that this is a huge misinterpretation.

The work of English philosopher, John Locke, is perhaps the most influential to Enlightenment thinkers such as Hume, Rousseau and Paine. Best known for his contributions to the liberal theories that moved many European nations away from monarchy to self-governance, Locke is also known as the originator of British empiricism. Locke's theory of the mind is often cited as the origin of modern conceptions of the identity, and particularly political identity and 'the self'. Although he did not have much to say about fathering *per se*, Locke's perspective on the mind as a *tabula rasae* containing little that was divinely given or innate was not only revolutionary but also suggested very specific relations between adult (males) and children.[5] Locke argues in *Some Thoughts Concerning Education* (1693) that children are not rational but they have the capacity to learn reason. They are on the way to becoming adults and the deployment of tutelage and education by fathers, in particular, moves them along an appropriate trajectory. If this may be seen as a framing of children's lives by adult males, then what Locke's work simultaneously opens is a critique of patriarchal fatherhood.

David Foster (1994) demonstrates that in much of his writing, and particularly in *Two Treatises of Government,* Locke's attack on divine pre-determination is also an attack on the patriarchal basis of fatherhood. Patriarchalism, at this time, justified itself through the notion of a natural and divine order: the law of god-the-father mirrored on earth as the law-of-the-father. Locke fundamentally disagreed with contemporaneous theological hermeneutics that placed the family as the foundation of political life and so his attack on patriarchalism prepares the way for his individualistic liberal politics. Locke distinguished political power from several other forms of power, and the most important of these is paternal power.

Locke's critique of patriarchy is scrutinized by a number of feminist scholars including Jean Elshtain (1990), but the focus of this attention is on gender inequities rather than on his treatment of fatherly authority and power. In considering these contexts, Foster argues that Locke's rejection of patriarchal fatherhood as divinely inspired is also a rejection of father's rule within the family. That said, Locke's differentiation between the 'private' family and the 'public' state creates a spatial frame, an artificial geography if you will, which forecloses upon the gender politics of families and leaves patriarchy intact (cf. Eisenstein 1981). Countering this, Foster points out that all Locke presupposed is that the relationship between fathers and children is a private matter, and this then leads him to the argument that education and moral teachings are wholly the responsibility of parents. Locke's point here, suggests Foster, is primarily to stop fathers from acting like kings, and vice versa. Foster argues from this that Locke does not simply separate parental (private) power from political (public) power while leaving patriarchy intact, but rather that the separation presupposes a different understanding of both. And importantly for what I want to argue here, this presupposition requires a reform of the patriarchal family and that reform is rooted in human emotions, and, in particular, desire.

---

5 In Chapter 2 of *The Geographies of Young People* (2001) I suggest that Locke's work might be considered – along with Rousseau – a harbinger of child-centred education and also a root of child developmental theory. For Locke, both knowledge and rationality are obtained incrementally and a large part of 'the self' is a conjoining of the naturalistic and the social.

Locke's primary desire is self-preservation, which he parleys into property; and his secondary desire is propagation, which he parleys into familial life. This simply designation was an attempt to upend Sir Robert Filmer's foundational *Patriarcha* (1680). Filmer was an English royalist who argued that all political authority and obligations – from kings to fathers – was derived from the fatherly power of Adam. For Locke to overcome Filmer's traditional notion of a divinely given law-of-the-father, he had to

> … make the family fully compatible with the primacy of the individual desire of self-preservation, and this has two main consequences. First, a concern for property comes to pervade all aspects of family life. Secondly, to make this possible, the relations of authority and obedience, which traditionally were considered the moral bedrock of the family, and which obligate parents and children to sacrifice property, liberty, and even life for the sake of family, must be loosened or undermined (Foster 1994, 647).

Locke's focus on parental desire leads him to advocate rewards, friendship or shaming as more appropriate disciplinary tools than beatings and harsh language (Locke 1693, 106–8). Locke believes that parents have power and authority over children to a certain age, but that power is conditional rather than natural. The problem with Locke's focus on familial desire revolves around men desiring only the act of procreation and not its consequences, what he calls "satisfying … present Appetite" (Locke 1690, 34; Foster 1994, 659–60). This clearly ties in with the arguments I raise in *Family Fantasies* concerning fathers' propensities to help out with, rather than take responsibility for, children.

This leads to an important question of priorities. Are fathers morally obligated to care for children even though such care may contravene their own private good? To this, Locke uses extreme examples of child sacrifice and cannibalism to point away from a fixed natural patriarchic morality and to suggest, rather, that there is no clear or automatic way to the preservation and nurture of children. Locke presents the family "… as a profoundly problematic institution that cannot, as patriachalists contend, supply a solid foundation or model for political order. The problem is not that … men are too possessive of children (and women), but rather that by nature men do not think of children (and women) as their own at all" (Foster 1994, 662). Given that fathers have passions for their children, Locke's solution is to guide men's fundamental desires towards the preservation, rearing, tutelage and education of children. Child-rearing for Locke, then, becomes an extension of the passion for self-preservation which, according to Foster's (1994, 662) reading, provides "the basis for a relatively lasting connection between fathers and their children without giving fathers the great powers patriarchalism thought necessary."

Hugely influenced by Locke, Scottish Enlightenment historian and philosopher David Hume is best known for reworking thinking about human nature away from divine determinism to a product of reason and insight. Presaging the basis of science for the next century and a half, Hume's understanding of the world is predicated upon radical scepticism, rational insight and empirical validation. But his conception of reason is not that of Descartes or even Locke. In his *A Treatise of Human Nature* (1739/1955), written when he was only 26, Hume counters the prevailing Western notion of reason by famously arguing both normatively and positively that "Reason is, and ought only to be the slave of the passions, and can never pretend to any other office than to serve and obey them … A passion is an original existence, or, if you will, modification of existence, and contains not any representative quality, which renders it a copy of any other existence or modification" (ibid., p 415). Arthur Herman (2001, 199) argues, albeit somewhat polemically, that Hume's "ought to be" in the above quote stood 2,000 years of philosophy on its head by pointing out that reason has a role that is purely instrumental in that it teaches us how to get what we want. What we want is determined by emotions.

According to Hume, the overriding force that guides action is not reason, or a sense of obligation to others, or an innate morality, but the most basic human passion of all, the desire for self-gratification (Herman 2001, 200). Herein lies Hume's understanding of the basis of self, family, and government. That said, Hume believed that passion was a problem, and as a social-conformitist he saw a solution to that problem in a society and a form of governance that propagated good habits.

As a propagator of good habits – like Locke before him – Hume lauds the family within the unique context of its history and geography, within a foundation of inherited material possessions, and through a lineage that is wholly male:

> … those, who can boast of the antiquity of their families, … that their ancestors for many generations have been uninterrupted proprietors of the same portion of land, and that their family has never chang'd its possessions, or been transplanted into any other country or province … when they can boast that these possessions have been transmitted thro' a descent compos'd entirely of male, and that the honours and fortune have never passed through any female (op cit., pp. 307–8).

Hume's radical scepticism of divine determinism placed him against patriarchy as it had been formulated by Locke's nemesis Robert Filmer. Hume argues that as an institutional form, the family models the beginning of the state in the sense that a basic union between parents and children is also the form of "a more numerous society." Parents' customs and habits – their rendering of justice and injustice – operate on "the tender minds" of children, and the exercise of authority is tempered by natural affection (op cit., p. 486).[6]

The part of Hume's work that bears most heavily on my current musings is not so much his focus on self-interest or how that translates into a patriarchal

6 A foundational context of Hume's treatise that moves his arguments away from those of Locke suggests that families are the antecedent to government and that both institutions are based on the protection of self-interests rather than divine intervention.

basis for family. Indeed, his speculations on the latter are not developed as fully as those of Locke. What is particularly intriguing for the strands of critical theory that I am trying to weave are Hume's elaboration of emotions over reason, and the suggestion – that permeates almost the whole of *A Treatise on Human Nature* – that the world is not really as it seems (Herman 2001, 261–2). In the words of Deleuze (2001, 35), who moves Hume's work away from the problematic 'emotions over reason' binary:

> David Hume pushes the furthest ... His empiricism is a sort of science fiction universe *avant la lettre*. As in science fiction, one has the impression of a fictive, foreign world, seen by other creatures, but also the presentiment that this world is already ours, and those creatures, ourselves.

Habit and convention – the foundation of Hume's notions of family and government – are beguiling because the nature of our world is uncertain, the society we create is ultimately unknowable, and our conclusions about our society and the world are almost always flawed.

With *Émile* (1762/1962), Jean-Jacques Rousseau was one of the first Enlightenment writers to acknowledge that children are of fundamental concern as different from adults. Children begin good, he argued, and it is only through 'man's meddling' that they become socialized into moral depravity. The story of *Émile* is based on a young boy's growth from infancy to manhood and how this development is directed wisely by a tutor. This kind of learning, accessible to only the most wealthy, generally required women to be at home and always available for their children. Rousseau uses this narrative to accuse affluent urban parents of attempting to replace sentiment with money by hiring paid intermediaries for their children, but at the same time he endorses a strict gender-role socialization: while Rousseau's Émile is educated as an independent and self-sufficient citizen, Sophie, the girl character in the book, is taught to be dependent and passive. In his contemporaneous novel, *La Novelle Héloise* (1761), Rousseau elaborates how households should be managed under the tutelage of a father; it presents a story for which the ideas were previously codified in his *Discourses on Political Economy*:

> [T]he husband should have the right to oversee the conduct of the wife, because it is important for him to make certain that the children he is forced to acknowledge and raise belong to no one but himself ... the children should obey the father, initially out of necessity and afterwards out of gratitude. After having their needs satisfied by him for half their lives, they ought to devote the other half to providing for his (Rousseau 1755/1988, 60–61).

It is not easy to reconcile Rousseau's support for paternal rule within families in these two novels with his insistence on political citizens in his later *Social Contract* (1763/1901). The differing scales of conduct are immiscible to the extent that some contemporary commentators argue that the distinctions drawn are not robust enough to justify his conclusion that, although all citizens should participate in legislative politics as equals, in families the father should command (Ritter and Bondanella 1988, 59, n. 2). Nonetheless, following Locke and unlike Hume, Rousseau saw the family and the state as separate but equally important realms in creating a moral society. Within this configuration, men were free, independent and self-sufficient citizens only if women and children were subordinated to male authority in the family. Clearly such a formulation contradicts his liberal promises of individualism, educability and achievement for all (cf. Eisenstein 1981). In his *Social Contract*, Rousseau makes a contentious claim that families in the state of nature lack fathers and suggests that attaching men to the more primary mother-child unit is a problem for civil society. His solution is the creation of the sentimental – affectionate and child-centred – family over the notion of the natural family.

Karen Struening (2001) notes that Rousseau's notion of the sentimental family raises huge questions concerning the relationship of men to families. Struening argues that Rousseau's model of the sentimental family – which grew out of his concern with integrating men into families – came to dominate Western European and North American notions of family. Rousseau's model pushed Locke's desire and Hume's passion beyond their mechanistic leanings to embrace emotional intimacy and love over economic and procreative functions. Taking on Locke's dilemma over men's commitment to children beyond their "sexual appetites", Rousseau argues that the family must be reformed so that it appeals to men. By so doing, he is critical of Locke's assertion that men who have no way of knowing if they are the father of a particular child are still likely to participate in the upbringing of the child. Struening (2001, 86) points out that Rousseau's reformulation of the family to appeal to men is nonetheless problematic because it promises them paternal certitude, power, prestige and pleasure. In this formulation, men and women who once lived independent in the state of nature, come to depend upon one another with men gaining the upper hand in terms of power. The upside of this formulation is that if men are taught to love the simple pleasures of family life, their erstwhile passions will be restrained and they will embrace both family and civic duties.

The Enlightenment writings of Locke, Hume and Rousseau are foundational to the creation of a family form and a spatial framing of that form in the private sphere with important connections to the workings of the state. It nonetheless seems clear that if these mechanistic functions exist, they are subservient to passion, desire and the sentimental family. It took a Frenchman travelling in the United States to make clear what the death of the patriarchal family looked like in the context of everyday life. Further, it took a Welshman working in the textile industry in Scotland to argue that the notions of individuation and property rights did not reside well with Rousseau's social contract. My story of the emotional work of fathering continues in the 19th century by moving away from heady Enlightenment ideas to the practical imaginaries of Alexis de Toqueville and Robert Owen.

Fathers are Citizens Like Everybody Else, Just Older and Richer

Following the Enlightenment thinkers, perhaps the most glaring portrait of 19th century fatherhood writ large comes from Alexis de Toqueville's famous rendering of everyday life in the evolving United States.

"[T]he distance which forever separated a father from his sons", Toqueville writes (1863, 233) in a chapter about the American family, "has been lessened; and that paternal authority, if not destroyed, is at least impaired. In America, the family in the Roman and aristocratic signification of the word, does not exist." By this, following Locke, and perhaps also Hume and certainly Rousseau, he meant that the patriarchal family – a family in which the husband and father had a large measure of political authority – is gone. Enlightenment thinking renders this family form obsolete. With the culmination of this thinking in the American and French revolutions and with the birth of modern democracies, the law of king and father is supplanted by "the long arm of government [that] reaches each particular man among the crowd separately to bend him to obedience to the common laws" (de Toqueville 1863, 234–5):

> The father foresees the limits of his authority … and when the time arrives, he surrenders without a struggle: the son looks forward to the exact period at which he will be his own master; and he enters upon his freedom without precipitation and without effort … In democracies, where the government picks out every individual singly from the mass to make him subservient to the general laws of the community, no [constituted ruler] is required: a father is there, in the eye of the law, only a member of the community, *older and richer than his sons* (emphasis added).

de Toqueville's assertions provide a clear link between the construction of private, propertied families and an evolving democratic state. The connection between individuation and private property is served well by the familial form that Locke, Hume and Rousseau preferred.

Whereas the work of Locke and Hume presage the theories of individuation and self-interest that help propel new liberal reforms in the 19th century and set the stage for Alexis de Toqueville's proclamation that modern democracy had defeated patriarchy, the work of Utopian thinker Robert Owen perhaps bears most heavily on thinking about families in communal rather than private, propertied ways. And, yet, with its push towards communal equality, Owen's Utopian experiment maintains fathers as "older and richer" and this was its ultimate downfall. In what follows, I spend some time reinvigorating Owen's ideas because I believe they offer an important signpost to the possibility of a "coming community" and fathers' "becoming other."

Figure 2.1 Robert Owen's mill and worker housing at New Lanark

An Attack on Individuation and Praise for Cooperation

A successful textile entrepreneur, Owen is perhaps best known for New Lanark, a pioneering industrial community outside of Glasgow. Whereas Hume's ideas were propelled by the way he constructed the desire to protect private property, Owen's crusade was to abolish it.

By 1799 New Lanark was the biggest cotton mill in Scotland, with over 2,500 people living and working in the village. Owen bought the enterprise in 1800 after a series of successful textile ventures in Manchester. He attracted world attention by the gradual introduction of his Utopian social experiment. Owen wrote *A New View of Society* in 1816 and used New Lanark as a testing ground for his humanitarian, social and educational ideas. Although he is best known for his Utopian vision of self-supporting communities, Owen's practical ventures in New

Figure 2.2 Workers' housing in New Lanark

Lanark and later in New Harmony, Indiana, focused squarely on improving local environments so that communities prospered. In opposition to doctrines of self-interest promoted by Locke and Hume, and the notions of social-conformitism as a control of depraved self-interest suggested by Hume, Owen (1816/1972, p. 22) nonetheless argued with Hume that emotions supersede reason and that it is possible, for example, to "*prevent* the existence of … crimes" through the principle of happiness:

> That principle is the happiness of self clearly understood and uniformly practised; which can only be attained by conduct that must promote the happiness of the whole community.

The idea of creating happiness through local environmental improvements, education for all and community cohesion permeates the whole of *A New View of Society*. In practice, starting with New Lanark, Owen attempted to improve the general environments by ameliorating working conditions and introducing better sanitation, education and housing.

Noted for his paternalism, the first problem Owen tackled was child labour. Presaging the later Factory Acts in the UK and Child Labor Acts in the US, he set the minimum age for employees at 10 years, he stopped the practice of

Figure 2.3 As part of his social experiment Owen reduced the space between worker housing, manager housing, the mill, educational facilities and other village functions

recruiting pauper apprentices and he reduced the hours of children's working day. His initial program for community development included providing housing, paving, sanitation and a water supply as well as building a company store that was cooperatively owned by the mill workers. He then extended his program to provide education for both adults and children.

A "new educational institution" was erected in the centre of New Lanark with an enclosed play area at its front. It is his writing on this subject that opens up Owen's sense of paternalism and makes a clear break from the individualistic/self-interest perspectives of early Enlightenment writers:

> The area is intended for a playground for the children of the villagers, from the time they can walk alone until they enter the school … As the happiness of man chiefly, if not altogether, depends on his own sentiments and habits, as well as those of individuals around him; and as any sentiments and habits may be given to all infants, it becomes of primary importance that those alone should be given to them which can contribute to their happiness. Each child therefore, on his entrance into the playground, is to be told in language which he can understand "he is never to injure his playfellows, but on the contrary he is to contribute all his power to make them happy." This simple precept, when comprehended in all its bearings, and the habits which will arise from its early adoption into practice, *if no counteracting principles shall be forced on the young mind*, will effectually

Figure 2.4 The recreated school at New Lanark, with the playground just outside the window

> supersede all the errors which have hitherto kept the world in ignorance and misery (op cit., p. 81).

Of course, much of Owen's perspective on education may be found, in more sophisticated form, in Rousseau's work. Owen's writing, however, emanates from the practicalities of new households in a new industrial system and is an important attempt to dissolve the geographic separation of the public and the private that was slowly evolving elsewhere (see Aitken 1998).

In terms of governance and civil society, Owen was ahead of his time in his support of women's suffrage. Nonetheless, all the documentary evidence supports the view that he imposed a strict discipline in the mills. If not patriarchal, he nonetheless locally acquired the reputation of a martinet. And although he promoted education for all, the tenor of learning differed for genders, with the "girls … taught to sew, cut out and make up useful family garments … to attend to the rotation in the public kitchen and eating rooms; to learn to prepare wholesome food in an economical manner, and to keep a house neat and well arranged" (op cit., p. 97).

Kolmerten (1990), in her study of Owenite communities in both the UK and in the US with an eye on gender ideology and power relations, suggests that this kind of inequality sowed the seeds of Owen's experiments' inevitable failure. She

Figure 2.5 Workers' housing in New Harmony
Picture by Ken Foote, used with kind permission

points out that although Owen's *A New View* promised gender equality there was "a glaring discrepancy between the public promises and the private realities of the communities" (op cit., p. 11). In New Harmony, Indiana (established in 1824), for example, women were allowed to vote, and boys and girls were ostensibly taught the same subjects, but in practice neither of these things happened. Married women and men were expected to work for the community, but only the women were expected to also care for their families. Female discontent was a large factor in the demise of New Harmony by 1828.

Community Economies

Despite practical failings, one of my main reasons for raising Owen's social experiment in some detail is the bearing it has on today's fathering practices and the possibility of a coming community. I am interested in the ways Owen's ideas and practices subverted both the insidious self-interested alienation of individuation and the public/private divide.

Gibson-Graham (2006) is concerned with the tensions between left labour politics and the kind of worker cooperatives that Owen championed and why those tensions seem to have little bearing for the contemporary dominant capitalist economic order.[7] They point out that common wisdom suggests a spatial framing of cooperatives as "individualistic, politically conservative, short-lived, labour-intensive, underfinanced and poorly managed" (op cit., p. 111 and p. 226, n. 21) and from this they note a penchant for essentialism.

> Essentialist ways of thinking constrict the ethical space of becoming, obscuring possibilities of (self)cultivation and the way that cooperative practice itself calls forth and constitutes its own subjects.

Of particular importance for Gibson-Graham (2006, 125) is that Owen believed that people could remake themselves through education as revolution and he believed that progress is made only if local environments and communities are conducive to education taking place. This can only happen when people live and work together.

For Gibson-Graham (2006, 88), a 'community economy' develops when there is a focus on an ethical praxis of being-in-common that cultivates awareness of (i) what is *necessary* to personal and social survival; (ii) how social *surplus* is appropriated and distributed; (iii) whether and how social surplus is to be produced and *consumed*; and (iv) how a *commons* (in the sense of a public good) is produced and sustained.[8]

7 Gibson-Graham (2006, 106-8) provide an insightful critique of Owen's Utopianism from Marx and others. They point out that issues of individualism and collectivism were traditionally seen by Marxists through the lens of class struggle where the collectivism of the workers was privileged over the individualism of the capitalists and entrepreneurs (op cit., p. 110). Within this context, the idea of community is only ever given naively. They point out that there is a complex politics of class versus community that is never fully developed in radical Marxist thinking, and elaborating this politics is a large part of their project. How this might relate to the dissolution of the private/public divide, and the demise of patriarchy in favour of the revival of fathering as a viable social and political practice is a large part of my project with the day-to-day fathering stories that pepper the chapters that follow.

8 Gibson-Graham use Owen's ideas as a springboard to talk about the economic community that has developed since the 1950s in Mondragón, located in the Basque region of Spain. Mondragón is exemplary as a successful cooperativist regional economy not just because of its ability to compete in a larger global capitalist market; importantly, the way it grows and distributes its surplus differs significantly from corporate capitalist ventures. While industrial cooperative growth is a priority, this goal is balanced by how the surplus is consumed locally. Social well-being is assured not only by paying adequate wages, but also through developing cooperatives (much like those suggested by Owen) in housing, healthcare, pensions, education and other social services. Gibson-Graham (2006, 125) argue that Mondragón's sustainable success is perhaps best exemplified through the ways it constructs communal subjects at a number of different levels – material, social, cultural, spiritual – but especially through ethical decision-making that encompasses individualism and collectivity. This ethic ostensibly began with a focus on youth-adult relations. As a Mondragón curate in the 1940s – years before the cooperative was established – Father Arizmendi practiced Owen's 'new view of society' through education programs, establishing youth, soccer and athletic clubs. He taught in the apprenticeship program of

The questions that this raises for me revolve around how parents (and fathers), and the family (as an institution) may be located as community economies.

Contingent Fathering and the Communal Household

> What ['a politics of the subject'] means to us minimally is a process of producing something beyond discursively enabled shifts in identity, something that takes into account the sensational and gravitational experience of embodiment, something that recognizes the motor and neural interface between self and world as the site of becoming both. If to change ourselves is to change our worlds, and the relation is reciprocal, then the project of history making is never a distant one but always right here, on the borders of our sensing, thinking, feeling, moving bodies (Gibson-Graham 2006, 127).

Discursive shifts around the concept of 'family' as a spatial frame suggest that it was writ large historically as *the* family, and problematically as the product of a single evolutionary trend (Glenn 1987, 349–50). It was never about a project of history 'right here' and 'embodied' in community, although it was hugely spatialized.

The problematic evolution of *the* family began with French philosopher Frédéric Le Play (1871) who argued that broad transformations from an extended family form in the middle-ages were changed to Rousseau's sentimental and spatially contained family form of modernity. By the mid-20th century sociologist George Peter Murdock was able to coin the term 'nuclear family' to describe "a small group characterized by common residence, economic cooperation, and reproduction, [including] adults of both sexes, at least two of whom maintain a socially approved sexual relationship, and one or more children, own or adopted, of the sexually co-habiting adults" (1949, 3). This functional definition of a family unit was derived in part from the earlier anthropological work of Malinowski (1913) who argued that the conjugal family is unassailable because it fulfils universal needs. The problem with this so-called evolution is that it is also an evolution away from extended functions, community and geographical contexts: the family moved towards autonomy (read powerlessness).

---

the largest company in town, Union Cerrajera, and set up an independent technical school. By so doing, he began to transform industry from within while at the same time he set up thousands of 'study circles' on social, humanist and religious topics. Gibson-Graham (2006, 126) note that Arizmendi was part of a larger movement in the 1940s and 1950s where a tradition of social Catholicism set out worker-priests to transform local capitalist enterprises through education and cooperative programs. They see this as part of a place-based global movement that shares a transformative social and economic vision that is ultimately pluralist, and that reflects a desire to proliferate ethically rather than to foreclose upon political possibilities. Gibson-Graham take Mondragón as their inspiration for an ethically based intentional community economy that is also sustainable.

Mark Poster (1978) was one of the first critical theorists to argue that this penchant for discursive framing was sorely inadequate because it could not pose important questions that rendered family change intelligible from perspectives that focused on embodiment and recognized a variety of contingencies. Theorizing from a functional perspective, he argued, rarely encompasses tensions in family composition, internal class configurations and the sexual division of labour, the evolution of the public and the private, and the spatial distancing of men, women and children. This kind of history and geography of the family finds its most eloquent expression with the work of Marxists, feminists and other critical and post-structural social theorists.

In what follows, I begin to look at these critical questions in the early modern period with a discussion of spatial framing and the creation and re-creation of families and communities as a consequence of the increasing powerlessness of families within emerging capitalist economies and the embodiment of fatherhood as an institution rather than fathering as a social practice. In Althusserian terms, the family and fatherhood become parts of an 'ideological state apparatus'.

The De- and Re-Territorialization of Families and Communities

Katharyne Mitchell and her colleagues (2004, 2) note that if most feminist geography on labour inside and outside the home has maintained the categorical binary of production and reproduction, the substantive distinctions between the two are consistently and productively blurred. Suzanne Mackenzie and Damaris Rose (1983) were amongst the first geographers to tackle the historic origins of the complex relations between social production, the circulation of commodities and the reproduction of labour power as they relate to spatial changes in families and communities. Mackenzie and Rose were intent upon probing the importance of spatial relations in the separation of the 'sphere of production' and the 'sphere of reproduction' with an understanding that neither capitalism nor work is so neat. They argue that the history of this 'separate sphere' has to be investigated in the context of people's struggles and contests around control over the means of production and subsistence because these struggles are integral to the development of the modern patriarchal family within industrial capitalism. They see capitalism creating the modern family form for its own ends and, as a consequence, the family is a historically contingent form ideologically contextualized by the state.

The viability of capitalism was predicated upon the need for a labour force which could be created out of the pre-industrial one, but that labour force was self-trained and self-regulated, and accustomed to an integration of living and working times and places. The new industrial labour force had to be disciplined, skilled in specific areas of industrial expertise, healthy and willing to work for a given number of hours every day. At first, there was an abundance of labour to serve the factory system in a relatively accepting manner but these people were mainly unskilled and unused to machine work. In time, the pre-capitalist household with its unity of production, reproduction and consumption collapsed leaving a void because there was no mechanism through which a skilled and disciplined labour force could be reproduced. Meanwhile, high infant mortality rates and unhealthy, disease ridden working-class residential areas left no doubt that the household was unable to reproduce a healthy and literate labour-force by itself. Not only was this a consideration for the expansion of capital itself, but workers – ill-fed, unhealthy and often homeless – were becoming more threatening as Enlightenment ideas of autonomy and individualism became popular.

By the late 19th century, as their fathers, husbands and brothers gained power and wealth in the public sphere, mothers, wives and sisters gained control over the domestic functions of consumption and became guardians of the reproduction of middle- and upper-class work ethics, at least in the limited sense of values because education and health were becoming increasingly controlled by the public sphere. An ethic of 'conspicuous consumption' developed with the tendency of wealthy patriarchs to demonstrate the ways they could 'keep' their women through numerous servants. As more and more elements of family service (education and health) and manufacturing (processing food or making clothes) were transferred away from the household, the home became recognized as a separate sphere in which the labour force could be indoctrinated with appropriate values and attitudes of discipline and service. For working-class women and children, domestic labour still was directed towards producing some goods for use within the family but these activities were increasingly being taken over by factories and institutions. A fundamental change was that domestic labour now had to reproduce workers and unless someone in the household sold their labour-power in exchange for a wage, the household would not be able to procure the basic needs for survival.

> Working-class women's work within the home thus became subordinated to the imperatives of the capital accumulation process, even though the forms in which many household tasks were carried on remained unchanged from the pre-capitalist period. Moreover, domestic workers now had to provide physical and emotional maintenance for wage-earning husbands who, unlike those in pre-capitalist society, had little or no control over their day-by-day, hour-by-hour working lives. The alienated workers would return home from work tired and very much in need of a haven (Mackenzie and Rose 1983, 164–5).

Spatial transformations and reframings such as the geographic imaginary of public and private spheres coincided with the creation of surplus-value and the setting in motion of a self-expanding capitalism which creates (and recreates) (i) the production of surplus-value in addition to value, (ii) the accumulation of profit and its reinvestment for future profit and, (iii) the continual separation of workers from what they produce (Mackenzie and Rose 1983, 161). Swerdlow and her colleagues (1989) argue that the patriarchal pre-industrial household lost a productive function that was intimately connected to community production and local geography and this constituted an important change that heralded the creation of an autonomous modern family form.

Gibson-Graham (1996) insert theories of class (and class exploitation) into this discussion of households and families so as to make visible the production, appropriation and distribution of surplus (labour) that goes on at a particular site. Imagining the working family as a unified class leaves unaccounted the notion of class within the family; that individual workers may have interests in dominating women and children (Poster 1978, xviii). This point is missed by viewing the family as a unitary phenomenon, unchanging in its life-cycle, and undergoing some kind of linear transformation through time. What is also missing, Gibson-Graham (1996, 65–6) go on to point out, is an understanding of what was going on in the so called private familial sphere as:

> … an autonomous site of production in its own right in which various class processes enacted … The (white) heterosexual household in industrial social formations has often been a locus of what we have called a feudal domestic class process …, in which a woman produces surplus labour in the form of use values that considerably exceed what she would produce if she were living by herself. When her partner eats his meals, showers in a clean bathroom, and puts on ironed clothes, he is appropriating her labour in use value form. Throughout much of the 20th century, this form of exploitation has seemed fair and appropriate because the man generally worked outside the household to procure the cash income that was viewed as the principle condition of existence of household maintenance.

The nascent discursive power of a spatial framing that separates domestic and public spheres lends credence to the notion that the disempowerment of family space (and women and children) is achieved through distancing from the means of production.

But what happens if we reinsert the family as a centre of production? For one thing, it takes apart the basis of the struggle to establish a new modern family ideal in upper-, middle- and working-class (white) families through the notion that fathers are needed to support their wives and children financially, and that the means of that support comes from a spatially separate public sphere. The powerlessness of the early industrial family, Marxists argue, arose in large part because the home and workplace became economically and physically, then socially and emotionally entrenched in separate spheres. The problem of powerlessness in modern families for Gibson-Graham (1996) is quite different, arising from an immiscibility between what was prescribed by prevailing gender ideologies to be evolving in the public and private spheres, and what actually took place.

Crisis Tendencies

As more and more of the responsibilities that had befallen families were transferred to the factory or public institutions, women brought their traditional activities into the wage sector. By the beginning of the 20th century the idea of women in the public sphere combined with declining marriage and fertility rates to suggest a culpable breakdown of *the* family and what Connell (1995, 65) calls the beginning of a crisis in naturalized "onto-formative gender practices." Families were still needed as a basic unit of consumption and reproduction of labour, but the role of women in the public sphere was increasingly seen as unnatural and even dangerous.[9] Being a 'father' became quite different from being a 'mother', and the roles of sons and daughters came to depend upon age, sex and consanguinity in inflexible and problematic ways (Bernardes 1985, 281).

The mid-part of the 20th century is cast by many as a time when the promises of the conjugal family in a system of modern welfare falls apart. John Steinbeck (1939, 3–4, 34, 35) describes well the fragility of this system and the faltering of the role of fathers in families as the capitalist enterprise breaks down at the local level:

> Men stood by their fences and looked out at the ruined corn. And the women came out of their houses to stand beside their men – to feel whether this time the men would break. The women studied the men's faces secretly, for the corn could go, as long as something else remained ... After a while the faces of the watching men lost their bemused perplexity and became hard and angry and resistant. Then the women knew that they were safe and there was no break.
>
> The owners of the land came onto the land, or more often a spokesman for the owners came ... The tenant men stood beside the cars for a while, and then squatted on their hams and found sticks with which to mark the dust ... The women and the children watched their men talking to the owner men. They were silent.

Following 60 years after Steinbeck, when Susan Faludi (1999, 595) talks of the "betrayal of the American man," she describes a post WWII masculine landscape littered with the broken promises of the "patriarchal bargain",[10] and a crisis of masculinity that goes well beyond economic insecurity:

> The outer layer of the masculinity crisis, men's loss of economic authority, was most evident in the recessionary winds of the early nineties, as the devastation

9 Mackenzie and Rose (1983) argue that organized labour's agitation for the "family wage" included concerns to keep married women in the home so that they could take care of the male wage-earner and his children. They go on to point out that the 'family wage' was a myth because working-class women almost always had to contribute financially to the household income. Most working-class women were employed in factories and the service sector, but for others an informal economy developed around activities such as sewing and mending clothes, or being paid by wealthier families to 'take-in wash' or 'child-mind'.

10 The patriarchal bargain is an ironic term coined by Deniz Kandiyotti (1988) to suggest what women and men got out of patriarchy.

> of male unemployment grew ever fiercer. The role of family breadwinner was plainly being undermined by economic forces that spat men back into a treacherous job market during corporate 'consolidations' and downsizing … As the economy recovered, the male crisis did not, and it became apparent that whatever men's afflictions were, they could not be gauged solely through graphs from the Bureau of Labor Statistics. Underlying their economic well-being was another layer of social and symbolic understanding between men, a tacit compact undergirding not only male employment but the whole connection between men and the public domain.

As men's loyalty to the public sphere was tested, their connection to the private sphere was eroded further. From the mid-20th century onwards there was, in Anglo-America, an increasing emphasis on the fundamental importance of mother-child relations, underscored specifically by the academic writings of Freud, Winnicott and Bowlby and the rise of self-help experts and pop-psychologists such as Benjamin Spock. When Freud called the mother "the first love object" (1900/1978, 369–70), he legitimized in experts from Winnicott to Spock the idea of women as the primary nurturers of children and men who, with the reduction of their real wage earning potential, were discouraged from a balanced work/family role (Mintz 1997). The paradox, of course, is that the emotional separation of men from families and the focus on property rights and individuation that grew from the Enlightenment does not reflect day-to-day life in the same way that it buttresses a set of hugely problematic societal moral values.

During the last quarter of the 20th century, women's exploitation in the domestic realm was seen as increasingly unfair and something to be struggled against. In part spurred by second-wave feminism and women's increased involvement in waged labour outside of the home, many argued for a more equal domestic load between mothers and fathers. Of particular importance in this debate was Arlie Hochshild's (1989) demonstration of women's "second shift" in the home after working in the public sphere. For Gibson-Graham (1996, 67), this was the beginning of an ongoing crisis of feudal domestic class processes. They note a move forward away from the feudal household to the notion of a communal household where all members perform surplus labour and jointly appropriate it. These new family forms are identified by Judith Stacey (1990) in her famous ethnography of working class families in California threatened by global inequities and shifts. Men who take on more domestic responsibilities may confront the loss of public and private status; women confront the mixed emotions associated with relinquishing more of the care of children.

In the 1990s, academic research – primarily from sociology and family studies – raised concerns about the marginalization of men in families. In *Fatherless America,* David Blankenhorn (1995, 67) famously critiques what he sees as a move towards labelling fathers as superfluous. He argues that it is a grave mistake to urge fathers to be more like mothers, and that fathers need to be more specifically masculine. Blankenhorn (1995, 67) lambastes expert discourses that suggest "social progress depends largely upon a transformation of fatherhood based on the ideal of gender role convergence." He goes on to argue that there is a problematic link between the cultural ideal disassociating fatherhood from masculinity and the rise of fatherless families. With a similar set of arguments, David Popenoe (1996, 169) suggests that men do not have the same "natural impulse to nurture" as women, who are "biologically more attuned to infant care." He argues forcefully that men are biologically inclined to play the role of "protector, provider, teacher and authority figure." Popenoe (1993, 538) claims that the new "streamlined" family can now focus upon its two most important functions: childrearing, and the provision to its members of affection and companionship. Accordingly, the family becomes the emotional centre of social life. Popenoe goes on to note that the 'family' is by far the best institution to carry out emotional functions. His arguments are compelling in the sense that he notes the importance of childrearing in a nurturing environment, but they are also grounded in modernist notions of neo-liberal individualism wherein the freedom of the individual subject is the paramount value.[11]

Alternatively, Gibson-Graham (1996) note that the communal household with equal, co-existing partners is not generalizable and is replete with a myriad of differentiated class antagonisms and conflicts. Stacey's "brave new families" – including communal households and solo households (cf. Pryor and Rodgers 2001) – may be seen as the outcome of struggles against patriarchy and gender oppression as well as the outcome of global economic restructuring. Gibson-Graham (1996, 68) see families/households as the outcomes of struggles around class, with three distinctive elements. First, rather than being governed by one spatial frame such as patriarchy or conjugality, families are a social site in which a wide variety of class, gender, racial, sexual, adultist and other practices intersect and second, because of this, it can be theorized as a locus of difference and constant change and, importantly "[e]ach local instance is constituted complexly and specifically, unconstrained by a genetic narrative or patterns from which it may only problematically stray."

Ideological Fatigue

Nearly two decades ago, Thomas Laqueur (1988, 155) lamented that "... we lack a history of fatherhood, a silence [that is] a systemic pathology in our understanding

---

11 Karen Struening (2000) points out that there is a strong relationship between Popenoe's notion of an affection-based family and Rousseau's sentimental family. In comparing Rousseau's ideas with those of Popenoe (1996), she shows that both 18th and late 20th century strategies for resolving the problem of fatherless families require a gendered division of labour that preserves and promotes gender inequality. She notes that while contemporary advocates of the gender-structured family, like Popenoe, make some concessions to the feminist demand for greater gender equality, they are in truth part of a long tradition that bases its ideal of family on the subordination of women to men.

of what being a man and being a father entail." It seems that man-as-father is either hidden or subsumed under the history and geography of man as a public figure and, unfortunately, the pervasive patriarchy that often accompanies this discursive framework. This is not only a hidden history of fathering, it is also a hidden geography, which is tucked away in the recesses of private life in much the same way that, until recently, the geography of mothering and reproduction rested fitfully beneath the 'dominant fiction' of public, male ideologies. The patriarchal bargain, which seemingly gave men power in a public imaginary not only preempted the idea of man-as-father – an embodied, feeling, sensational and moving subjectivity – but, also, ultimately, betrayed men (and women) and removed from reach fundamental promises of the bargain such as security, a family wage, and the prospect of raising healthy and educated children.

Kaja Silverman (1992, 15) points out this "… 'dominant fiction' or ideological 'reality' solicits our faith above all else in the unity of the family and the adequacy of the male subject." What she means by this is that, at least in the common wisdom of Anglo-American notions of well-being, there is a relatively unwavering confidence in the family as an institution, and that structure comprises a male subject who is able to recognize himself as father and what he does as fathering. And yet, within the discursive frameworks that define contemporary families and as part of what has been called the crisis of masculinity, Silverman (1996, 16) goes on to point out, "the prototypical male subject is unable to recognize 'himself' within its conjuration of masculine sufficiency" and from this discursive framework "our society suffers from a profound sense of 'ideological fatigue'."

This chapter has sketched a rudimentary geography of the ideological fatigue of the father-subject-self that suggests both failure and promise. The failure emanates from frames that constrained families to particular spatialities and place men at the margins of an imagined private domesticity. Along the way – beginning with Enlightenment thinking, continuing with Owen's *New View* and Gibson-Graham's *Post-capitalist Politics* – there are glimpses of promise and hope from a more open communal household and a community economics less tied to individuated, profit- and property-focused forms of capitalism. The chapter brings us to a consideration of the neo-liberal subject in relation to the current regime of production/reproduction/accumulation and in relation to state apparatuses. In order to understand how fathering as a work is changing today, it is important to know more about how individuals make and understand themselves as fathers, and how these subject positions are constituted spatially through the discourses of public and private, inside and outside, home and away, imaginary and real, and so forth (Mitchell et al. 2004, 3).

Chapter 3
Recovering Fatherhood

Ed showed up in Chapter 1. He helped me through the issue I was struggling with regarding memory. Using an example of a memory about playing with a coat-hanger as a child, he pointed out that although memories are hugely contrived they are nonetheless authenticated through emotion. The memories are not real in the sense, as with the coat-hanger, that the events may not have happened to you, but the emotions that encompass the memories are incontrovertible. During our supper that night, my conversation with Ed turns to the ways we get caught up in categories. Unlike the affective component of memories, he sees categories of meaning as hugely controvertible.

"I totally agree that men are as intimate and communicative [as women] and all of the stereotypes about them are so false, and the stereotypes about women are also equally false." Ed thrusts his fork in my direction. "You know it just says that we're kind of caught up [in categories], I don't know what that is about and I don't know that [the reality] has to be broken so much as just revealed. And I don't know that people don't really know that that is not true at some level. Examples … and stories like that man in the book, those are so …, that is such a beautiful way of allowing people to experience the differences without being hit over the head or getting into arguments about them, or getting clinical, you know?"

"The man in the book?" I ask, wondering if Ed's soliloquy on fathering as a category of existence will survive my interruption.

It does. Beautifully.

"I sent you a book one time, did you read the book?[1] Do you remember the [story] where the wife showed up? They were out in the country somewhere, I don't know where it was, but the husband and son were fishing and the wife showed up and just watched them. And she watched them for a long, long time and never said a word. Even to her there was so clearly a bond there, and that so typifies one of the differences."

"The silences?"

"Yeah, and sometimes it can be uncomfortable. It doesn't matter. And sometimes, and a lot of times, it is the inability to express things that is important. But I mean you know that doesn't mean that you have to fill that discomfort with *useless chatter*." Ed slows his speech down to emphasize the last two words and then silently cuts up the steak on his plate.

---

1 Ed sent me a copy of Kent Nerburn's (1999) *Letters to My Son: A Father's Wisdom on Manhood, Life and Wisdom* (Novato, CA: New World Library).

I enjoy Ed's company, and I revel in the silences. We sometimes meet for breakfast on Saturdays with a group of men to share food and to chatter *usefully* about things that bother. For me, Ed is part of my *coming community*: an emotionally connected collective that is part family, part community, part therapeutic support group. Sometimes our chatter leads to silence, sometimes to tears or laughter, sometimes to sadness or anger. Over time a catharsis of sorts evolves from these shared experiences. We learn from each other; we trust. I've changed a lot since my son was born 18 years ago and what happens in groups like this is a large part of that change, although I am not sure what precisely I change towards. Perhaps I grow up, perhaps I gain wisdom from others' experiences. It doesn't matter. The *coming community* is unfocused, comprising an "inessential commonality, a solidarity that in no way concerns an essence" (Agamben 1993, 19). Some of the communities of which I am a part do have a focus: I've spent a lot of time with men in focus groups that look at violence, racism and sexism; in group therapy sessions the focus is on interpersonal relations; in meetings of Alcoholics Anonymous the primary focus is on recovery from addiction, of which families are a huge part; on weekend retreats the focus is often on spirituality.

Prior to the birth of Ed's son, I sat around a fire circle on a cold Californian beach with 20 other men and we shared with him our experiences of fathering. I looked around at faces I'd seen at AA meetings, on weekend retreats and in coffee shops. These men are my coming community focused, just for that moment, on Ed's nascent fathering. Around the fire circle, we ceremoniously lift Ed and collectively cradled him in our arms, promising to support him on his fathering journey. At 52 years, Ed was concerned about the responsibility of a young life so late in his life. We cradle him, offering physical and emotional support and the knowledge that he need not make this journey alone. Seven years – and many Saturday breakfasts – later I sit with him in a restaurant to share a meal and tape-record his views on fathering.

The previous chapter sketched some histories and geographies of fatherhood and ended with ideological fatigue. It was not an attempt to re-etch or trace an identity of fathering but to disrobe the imaginaries of fatherhood writ large and to provide a mapping that is disconnected and thus, the possibility of re-connection in different ways. Ernesto Laclau (1990) talks of the importance of dislocation of identities/spatial frames to make space for the political. The identity of fatherhood cannot be seen as the property of a bounded and centred ideology that reveals itself through history or geography. The identity of fathering, as a work and a toil is open, incomplete, multiple and shifting. As Chantal Mouffe (1995) and other poststructuralist theorists note, identity is always partial, hybridized and nomadic. Identity is also about (and fundamentally different from) difference.[2] The importance of dislocation is that it makes something else – something different –

2 For Deleuze (1994, 40–41) identity is re-read as a 'second principle', "as a principle *become;* that it evolve around the Different: such would be the nature of a Copernican revolution which opens up the possibility of difference having its own concept, rather

possible, and for Mouffe (1992) one of those possibilities is a transformed political community.

As a move away from the spatial frames and ideological fatigue of the previous chapter, this chapter is a mapping of fathering as fluid, open and vulnerable. Laclau's project with Mouffe is to find, rethink and dislocate the identity of 'the social' as something that is not closed by a structure (Laclau and Mouffe 2001). Similarly, Judith Stacey (1990) attempts to dislocate the monolithic family form with the notion of complex family structures that respond to economic conditions, and Gibson-Graham (1996) take on the structure of the economy writ large as capitalism to suggest the opening of economic communities. As this fathering map project unfolds, my interest is to do the same thing with the awkward spaces of fathering: to re-imagine the subject and subjection of fathers as unbound and fluid, simultaneously open and filling, at the same time an illness and also a remedy.

To a large extent the non-kin-based family gatherings that are part of my support system are about trust and vulnerability (Gibson-Graham 2006, xii). In her essay on transgression and transformation, bell hooks (1996, 20) argues that "there are moments when submission is a gesture of agency and power." There is a distinction, she points out, between conscious surrender as an act of choice and the submission of someone who is victimized and without choice. At times, as a man and father, I surrender to the wisdom of others. Within the forlorn vestiges of a patriarchal culture that creates, at least in part, the categories of woman and man – of father and mother – it is important to know the difference between surrender as an empowered choice and surrender as a consequence of victimization. The former, notes hooks (1996, 21) is a powerful form of self actualization: "To love fully one must be able to surrender – to give up control. If we are to know love, then we cannot escape the practice of surrender." Within the ideological fatigue that is patriarchy, the men who are part of the groups I frequent sometimes take a risk with surrender. By so doing, they create a space of vulnerability where they are open to wounding, but it is also a space of healing and transformation. With a recognition of prior wounding, I've seen men move beside their victim to own their actions/doings as fathers.

I've joined men in other forms of gathering. I've participated in academic sessions at international conferences where the topic is geographies of children in relation to fathers and other caregivers. I've participated with young and old men in a variety of recreational activities: biking, hiking, and kayaking. We join to eat breakfast, lunch or dinner. These encounters are all part of my becoming story, my gathering of other men's stories, my connections, my communities, my fathering.

In what follows in this chapter I focus on the contrivances and co-creation of emotions and fathering practices with intent in the chapters that follow to say something about how masculinities are differentiated spatially. To begin, I offer a brief conversation with Benjamin that focuses on his emotional connection

---

than being maintained under the domination of a concept in general already understood as identical."

with his daughters and the vulnerability he felt when his estranged wife fought to remove them from him. This is followed with a fuller discussion on the context of differentiated fathering with intent to elaborate the multi-faceted complexity of representations that are, following Deleuze (1986), best articulated from non-representational standpoints that focus on affect. As I hope to show, affective stories not only invite visceral reactions that shock, they also say something important about space and difference. I close the chapter with some discussion of how these embodied reactions play to larger contexts that situate the possibility of social transformations. This provides a conduit to the visual thumbnail-sketch experiments that are contained in Chapter 4.

The Emotional Work of Fathering

I've known Benjamin for about eight years. Self-trained as a computer software developer, Benjamin works for a local web-based company. We met at a men's retreat and got together for a number of years on a weekly basis with some of the other men who were part of the retreat. The weekly meetings were intended to provide support with day-to-day pressures and larger anxieties as they arose. The weekend's focus on Jungian shadows was also the basis of therapeutic techniques employed at our weekly gatherings. We used a variety of experimental techniques learnt on the weekend and other similar workshops to tease out the ways that our current emotional experiences were rooted in past experiences that we do not see clearly (hence, shadows). The goal was to shed light on the shadow to the extent that its force might diminish. Through these weekly meetings I got to know Benjamin quite well and I shared in the losses and grief of his failed marriage and his struggle to maintain a connection with his daughters.

I switch on my audio recorder and ask Benjamin for his fathering story. I know that Benjamin is fairly well in touch with his emotions, but I am surprised at how quickly he gets to that place.

"Reflecting on all the emotional work I'd done, I judged I was ready to have kids," he says, "and it tied in with my wife's biological stuff where she was out of school and other logistical things, but for me it wasn't until then that I felt that I was ready to be a father and so we got married and the first go around, you know, with her cycle, we were successful."

I say: "Tell me about the emotional work that led you to a place where you were ready to be a father."

Benjamin pauses for a moment.

"I'm flashing back and forth to a lot of different times starting from being a teenager until my 20s and my 30s. Kids were important to me in my mind. It seemed like an improbable question to me … that I could ever be a father. I mean that is just the way I saw myself. When I was 19 I started going to see a therapist and, I didn't know why, I just knew stuff wasn't right. I am sure that was the beginning of my 'peeling of the onion'. So a lot of my early 20s [was about] becoming aware of the spiritual part of me. I started exploring that pretty intensively. It just kind of exploded one day. I had always been very left brained: you got to prove it me otherwise it doesn't exist scientifically. A new friend of mine who did astrology said 'let me do your chart', 'okay,' I say, 'whatever, sure'. And a week later she comes back and starts to tell me all these deep insightful things about me that I'd never talked about, certainly not with her. And at that point it opened me to the possibility of something else that I could not measure out there.

"So, you know I don't remember when it was that I wanted to have kids … I didn't feel ready. The idea of being a father at that time never entered my mind but in the future it was there. You know it is just that I have been doing a lot of different work in a lot of different ways on this all through since I was 19 and it has been onion peeling. So as that work continued I met my wife: I was 30 … I just wanted to become friends with this woman, even though I immediately felt attracted to her. We spent many, many an hour just talking and it was in that moment one night that I realized she would make a really great mother for my children. And, you know, flash forward 15–20 years later – whatever it is now … I knew she really would be a great mom, and she has been. And at that point the idea became a lot more cemented in my mind about kids and, and it was really a question then of logistics."

"So logistically you were ready and you conceived on the first try. Tell me about your emotions about learning you were going to be a father …" I wanted to get Benjamin back to talking about emotions. He did.

"Very excited! I remember going into the bathroom and looking at the kit, we have a picture of it." Benjamin laughs at the idea of a photograph of a positive pregnancy test. "It was really exciting. You know we both went 'aeyah aeyah', my god what have we done … we have done it haven't we? But it was, yeah, we've done it and we were both very excited. I was excited."

"So, what happened over the next nine months, can you remember?" I ask. Benjamin's left-brain logistics kick in.

"Yeah, we learned a lot, we educated ourselves: she definitely was in the lead in terms of research and stuff but I was right there. She'd bring home a book and I'd be all over it. And you know in my mind it was a chance to do it right, to be present for my children, to be loving, you know I had whacked away a good chunk of the onion by that point. Still a lot more to go, but I felt emotionally ready. Part of that was also just being emotionally present for my kids, being able to give them a vocabulary which I never received and to …"

"What do you mean by that? Vocabulary?" I interrupt. What I get in return is quite startling.

"To be able to: I never ever, ever was able to express any of my emotions. Not only was it frowned upon but it was never taught: this is what sadness feels like and this is how we handle it, this is what we do when we get angry; it is okay to be afraid, this is how we handle fear. That was never shared with me and that was one of my biggest conditions of being a father was to provide that for my kids."

What I learn from the men I talk to about fathering is that emotions are important and become more so as they develop a vocabulary to talk about them. For Benjamin, the emotional vocabulary comes from his work in therapy and at men's retreats.

My speculations on the question on emotions and the work of fathering embrace post-structural feminism and are inspired by the early writing of Gilles Deleuze (1971, 1986), with particular emphasis on what he says about affect and difference. I am also intrigued by the space of those emotions, because I believe that some of the spatial frames elaborated in Chapter 2 closed down men to not only their emotions but also the capacity for a politics of becoming something different.

Deleuze's concerns are spatialized in his work with Félix Guattari (1988a and 1988b). The corpus of Deleuze and Guattari's work suggests a non-representational and non-discursive way of knowing that relates the production of difference to ways that ordered spaces are disrupted in quirky, emotive, and relational ways. Here, I talk about the possibility of vulnerability and affective disruptions, and how those play out in the work of fathering; and how that play moves me towards a coming community. I focus only on a few fathering stories in this book, but I nonetheless highlight multiple masculinities that resist, are one with, and are differentiated from the hegemonic patriarchal forms of the previous chapter. My intent here is not to resolve and subvert patriarchal ways of knowing, but to raise other forms of fathering that are about love and vulnerability rather that rules and control, that are about communication and empathy rather than individuation and property rights. Seen in these multiple ways, fathering may be collapsed into Deleuzian *illnesses* and *remedies*.[3] What I want to do here is spin this two-sided Janus-coin so as to blur the edges and features that define it.

Up Close and Personal

"So what were your feelings and emotions around the birth?" I ask Benjamin.

"Ah, they were mixed," he says. "Being present and being there really was good and at one point during her pushing, you know I just started crying. I was really … it takes me to tears now, watching my daughter's head come out."

A long pause. Benjamin's eyes get glassy.

"Yeah … I've had a handful of miracles in my life and that was the second one, the first was my own birth. Yeah so that was way fucking cool, it was really great."

Benjamin's laughter is hugely joyous.

3 In the same sense that Deleuze's work speaks to double articulations between molecular and molar, and content and expression, so too it articulates a simultaneous sense of illness and remedy in an action depending upon which level it is encountered (see Aitken 2007).

"They started to hand my daughter to me," he laughs, "I say to the nurse 'I think you should really give her to her, she was the one who did all this work'. That was the right call and then my wife held her for like 15 minutes and then passed out. I got to hold Rebecca for three hours."

"How did that feel?" I ask. Seconds pass. By the end of a minute tears are flowing down Benjamin's cheeks.

"I really can't talk about it."

"Yeah." I respond sympathetically. Tears are in my eyes also.

"Three of the most special hours I've had in my life."

More tears.

"Yeah. It was really a treat, you know?"

I did.

"I made sure I made time with my other daughter when she was born but it wasn't the same, it couldn't be. It was ... it was ... just. There was just nobody there. We had a friend who was in town who came in very briefly and stayed in there with and ... and I think he went and brought back food."

"Three hours!"

"And I just sat with her."

A bonding. An embodied connection. An intimate space between a father and a newborn child. A memory, and a continued overflow of emotions. What precisely is the affective mechanism that engages me and Benjamin? What is its lasting effect? How does it create Benjamin and me as fathers? These are heady questions that I intend to spend most of this chapter tip-toeing around because I think I do the emotions an injustice with too much analysis. Moreover, to delve into their depths assumes, I think, an unhealthy penchant for universality when what I really want to talk about is difference.

Issues of difference arise with Deleuze's theorizing, which I engage more fully in the latter half of this chapter. What I want to point out at the moment is that affect is about the virtual, and it is as powerfully embodied and visceral. As Massumi (2002, 30) points out, "something that happens too quickly to have happened, actually, is *virtual*. The body is as immediately virtual as it is actual." Forms and meanings of fatherhood prescribed in the previous chapter and outlined by my beginning conversation with Ed in this chapter point to the unrepresentable, to that virtual dimension of affect before images are registered by conscious thought (McCormack 2003, 495). This is a poetics of space that Bachelard (1958) was unable to register in his phenomenology of consciousness. The central issue of difference, I think, arises from this pre-conscious connection. It is consequently important to examine how conscious categories not only create borders, exclusions and dichotomies, but how those processes are different when they begin with embodied affect. When I cry with Benjamin, it is about communion and empathy, and it is about surrender. Difference is placed elsewhere.

Deleuze (1986, ix) suggests that our emotions present to us a "pre-verbal intelligible content" that is not about linguistically based semiotics, a universal language, or some existential or Lacanian lack. Affection "surges in the centre of indetermination" between the perceptive and the active, occupying it "without filling it in or filling it up" (Deleuze 1986, 65). Affection re-establishes the relation between "received movement" as perception moves me from the total objective to the subjective and "executed movement" when I grasp the possibility of action. "It is not surprising," Deleuze (1986, 66) goes on to note, "that, in the image that we are, it is in the face, with its relative immobility and its receptive organs, which brings to light these movements of expression while they remain most frequently buried in the rest of the body." Summarizing a number of writers who are influenced by Deleuze's work, Christopher Harker (2004) points out that affect is about an intensity that exceeds representation. I look into Benjamin's face and I see his tears, and I am with him holding a newborn.

Incommensurable Differences

Theoretically, it is worthwhile looking at the ways emotions and affects differ. Whereas it is not difficult to write about emotions it is difficult to describe affects. Emotions are a subjective consensual understanding about affect; they are the part of affect that is owned and recognized. I know faces that are angry, sad, fearful or joyous and I understand the emotions that they represent. I cry with Benjamin as his story touches the parts of my fathering that I hold dear. But the totality of affect is something more, something that goes beyond moments of joined emotions. Massumi (2002, 22) argues that affect is about emotional intensity that is not directly accessible to experience and yet is not exactly outside of experience either. This elusive intensity is felt in my body rather than understood in my mind. Gut-wrenching experiences tighten my jaw and stomach, or weaken my legs and bladder. These are visceral, embodied reactions that cannot be reduced to simple expressions of joy, fear or anger.

Benjamin and I talk some more about the emotions we felt when our children were first born; rocking them in the hospital, the fearful journey home and the sleeplessness of the first few nights. Our shared joy continues until he gets to the part of the story where different emotions arise. I ask Benjamin about his community of support with his wife and child now at home.

"So, em, what kind of support network did you have, if any, at that time? For being a parent?" I ask.

"Well," Benjamin pauses, something else is going on for him, "three days after …" he pauses again and then sighs. "Awe fuck …" There is a look in Benjamin's eyes that reflect a different range of emotions. "For a year and a half before [my daughter] was born my mom was battling cancer and …"

More tears, but the emotion is different. Anger.

"She actually had surgery in Tijuana because the American system wouldn't do it for her. And the surgery went really well and she would come down every couple of months and get checked out by the doctors, great guys. And, em, like 14 months later they found evidence that some of the cells had migrated from her lungs to her, her … there was some growth coming back. So, I think it was February … two, three days after Rebecca was born my mom came down to begin experimental treatment because at that point they had discovered that the cancer … no they were going to do an experimental, they were going to put a shunt in her but she was going to have to live down in Tijuana and see them regularly.

"A piece of history at this point: My ex-wife and my mom got along great before we became romantically involved, and then it all changed. In the late 80s we were living in this house as housemates and my mom would come up to visit me, and they would yack for hours and hours and hours and hours. But all that changed when we got together romantically. My take is that they were so much alike that my ex-wife couldn't stand what she saw in herself and took it out on my mom.

"But it's now a few days after Rebecca's birth, and my mom has come down to begin her treatment in Mexico the next day. We are all in the living room together, and I am sitting in the other chair, and before long they proceed to go at each other like lionesses, fangs bared, claws out. I felt my nuts smoking between my legs, it was so fucking horrible, it was so horrible and …"

"How soon after the birth?" I am shocked.

"This is three days later, one, two, three days later. I wanted to fucking die, I couldn't believe it. I knew my loyalties lay with my wife but my mom was dying and, and my wife was just being way over the top … I was spending my time going down to Tijuana a couple of times a week, working 40 or 50 hours a week, taking care of my wife, cooking, doing the laundry, doing all that stuff. And so it was a bitter-sweet time."

I get a sense of Benjamin's communal household and the ways he feels buffeted between his mother, his wife and his new child. What ways does difference show up in the household? If this is a time of huge emotional upheaval for Benjamin, then it is important for me to embrace the power of non-representational ways of knowing, but where does this leave me with regard to difference? The use of non-representational theory suggests a valuable lens on a more mobile form of subjectivity, but Benjamin seems very much stuck in a caretaking role. Deleuze helps here in terms of understanding an ontological move towards the domestication of difference and the construction of identity.

For all the power invested in representations of men as patriarch, they are inadequate for ontological reasons because differentiation is explained by reference to a discrete set of variables (e.g. having a phallus or not), thus encasing identity within a dialectical opposition of presence or absence. In speculating on the social construction of nature, David Lulka (2004) argues that inadequacies such as these pertain primarily to the propensity to diminish the presence of differentiation in dominant perspectives. Deleuze (1994) is correct in stating that representations based on opposition are fundamentally flawed because they subordinate difference to identity. For Lulka, the act of representation is really about the "domestication of difference" (cf. Carrier, 1998, 189). I'll have more to say about domestication in Chapter 10, but the questions that this raises for me here revolve around the ways that Deleuzian affection-images such as the image of Benjamin stuck in the middle between his wife and mother, enable resistance to hegemonic norms and the celebration of difference. Massumi (2002) argues that taking a Deleuzian perspective on affect begins with "shocking my thought" because this suggests a heightened awareness from which difference may foment. But how exactly does that occur?

"What was the argument between your wife and your mom about?" I ask. I am still shocked. And maybe it doesn't matter. Benjamin's emotional responses – today as then – and his memories are his truth.

"I couldn't even tell you." It does not matter, this is not about the accuracy of his memories. It is about his emotional response. "My wife didn't like how much [my mom] was leaning on me, and she thought she was using me. And, you know there is probably some truth or value or whatever in all that, but …"

"You were stuck in the middle."

"I was stuck in the middle. I mean I didn't have any … the only way I could deal with it was to serve both of them as best I could and it was a losing situation either way. Things were bad for a while."

"In what sense, bad?"

"Oh, the anger that lingered. My mom was finally … the cancer spread to other organs and then that was it. She went back to Santa Cruz and I split my time between the two places starting in the end of May until she died near the end of July. So for two months I was flying back and forth like every other week or so. And I'd stock up on food or I'd cook food at home and have it ready for my wife. I'd do the best that I could and her accusation was: 'You abandoned me. You abandoned me and my daughter.' And that, oh god that hurt so badly. Even [my wife's] mother was just horrified that she had done that … her mother had taken care of my wife's grandmother through cancer and she knew what it took."

If Lulka is right about the domestication of difference, then does a similar process of domestication occur on the representation of men's emotions and how is this elaborated through Benjamin's memories? The answer, I think, lies with hook's (1996) notion of vulnerability and love that I raised earlier in the chapter, but I cannot get back to that discussion before I more fully think through what Deleuze might have to say about affect and difference in the emotional work of fathering.

The Domestication of Difference and the Non-Representation of Fathers

With his criticism of representational theories, his focus on affect, his relevance to theories of the body, and his writing on the centrality of spatial metaphors, Deleuze's work reinvigorates writing on embodied, emotional geographies and the celebration of difference. His work enables social transformations to the extent that it opens a door for a reappraisal of the ways fathers are represented. A core criticism of representations is their lack of connections to the material world and for Deleuze (1986, 1988), in particular, it is their uselessness for understanding embodied affects. McCormack (2003, 493–6) points out that Deleuze reworks

Spinoza's cartography of affect in ways that are crucial for attending to the unrepresentable, and especially the body and emotion in geographical research. This reworking avoids treating emotion as the outward expressive representation of some inner subjective reality. It is a treatment that does not supplant or denigrate thinking. Rather, it extends the field in which thinking emerges by making more of those affective capacities and bodily reactions that are less representational. Harker (2004) argues that with Deleuze's insight, a new practical-theoretical grammar is produced; one which recognizes the roles that embodiment and emotions play in our lives, and which does so without colonizing or claiming to fully represent those roles.

Deleuzian theory suggests a new set of relations with the material world, one which challenges static representations of reality and identity that, in turn, facilitates a caricature of men embedded in their own power, a form of power that is always about the construction of dominant hierarchies. From this, Lulka (2004) makes a stunning conclusion: when encountering the representation of political identities – male, female, father, mother, animal, alien – their embodiment and disciplining effectively reinstates vertical hierarchies rather than generating ideas about as yet unrealized horizontal spatial relations of autonomy. It is to those horizontal spatial relations and their power that I want to get to, but first I want to take some time with Deleuze's notions of non-representation. To augment the differentiated capacities of men, representations of masculinity must be discarded in favour of approaches that foreground the embodied, visceral nature of existence, and encourage fluid affective relations (Smith, 1998). What is needed, I think, is a new way of thinking that affirms the ability of fathers to inhabit bodies and spaces in diverse ways, and also disavows the constraints imposed by modes of representation that highlight men as perpetrators or victims. In this way men are empowered because they are not viewed as active or passive; neither locked in a condition of representational stasis as stoic and in control or one which highlights the power of their victim. Yet even this specific form of reworking serves merely as an improvisation in redefining men's differentiation. In actuality, a further push towards a non-representational theory is ultimately needed to dissolve the hierarchy that currently separates fathers from children, men from women, and men from other men.

A spatial image from Deleuze and Guattari (1987, 131) that I particularly like and which seems to fit as a new option is the rhizome. Biologically, rhizomes are subterranean root-like plant stems that grow horizontally and autonomously, often thickened by deposits of reserve food material. They are distinguished from true roots because, like plants, they produce buds, nodes and scale-like leaves, and they produce roots below and shoots to the upper surface. For Deleuze and Guattari (1987, 131), they "strangle the roots and scramble the codes of all arboreal and sedimentary thought." They take to task my understanding of the ordered, hierarchical/Oedipal basis of father/child relations: "The rhizome is altogether different, *a map and not a tracing.* Make a map not a tracing" (Deleuze and Guatarri 1987, 12). For Deleuze (1986, 59), there are so many presentations of planes, and they correspond to a succession of movements in the universe. The material universe, what he calls the plane of immanence, is the ordering, fitting together, machine assemblage in a rhizomatic space of affective movement-images:

> We may therefore say that the plane of immanence or the plane of matter is: a set of movement-images; a collection of lines or figures of light; a series of blocs in space-time (Deleuze 1986, 61).

The task of critical theorizing is not the formation of conclusions or generalizations, but rather the pursuit of, and creation of, new concepts that, like the rhizome, are commensurate with the shifting character of the phenomena they investigate. By doing so, a new way of knowing unearths the cracks that suggest "new perspectives for living" (Lorraine, 1999, 203). The concepts and relations that Deleuze and Guattari present are composed of ceaseless intensities that jostle and erode striated spaces (such as the stoic spaces of patriarchy). These concepts are no longer viewed as fixed categories, but as fluid constructions and deconstructions.

In discussing rhizomic space as a product of resistance and its relation to identity, however, it is important not to overgeneralize or homogenize the process of dissent itself. To do so would deny difference in yet another way. Rather, resistance is fundamentally related to the current constitution of the body, and thus it is highly individuated and individuating. Individuation is exhibited by, and expressed through, forms of resistance that elaborate many kinds of spaces, encompassing numerous temporalities. Movement as a form of resistance, for example, includes not only broad horizontal shifts across relations, but also the vertical shifts through institutions. In addition to these distinctions, there are always moments of conformity. Men, at some times and in some places, are complicit with and project the rule of the father. Indeed, conformity is an important spatial aspect of Deleuzian theory, for to focus exclusively on resistance is an act of generalization and purification, effectively loosening and making more mobile the connections between being and becoming.

Marcus Doel (2000) outlines the dyad 'is/and' – which is similar to Deleuze and Guattari's distinction of being and becoming – to distinguish practices that are stagnant in nature from those that are fluid and tend toward dispersion. As Deleuze's theories suggest, the multiplicities that comprise fathers emerge from a succession of 'ands'. Deleuze and Guattari (1988) develop a multitude of concepts to express the propensity of material to mutate, transform, and thus give rise to a plurality of spatial formations. Among others, these include the rhizome, but also the assemblage, the line of flight and the becoming-other. And, as Lulka (2004) notes, it is the intervals, the spaces in between, the lines of flight, and the movements that remain critical because they do not embody transcendent qualities but rather the experience of existence. In each of these, a final destination will not be found and this is an important geography.

The call to understand emotional bodies is certainly a call to find the fluidity of neo-liberal subjectivity but it is also a call to understand its relations to spaces, places and contexts. For Benjamin it is stuck between his mother and wife, it is on the way to Tijuana or Santa Cruz, it is closeness to his daughter. My belief is that as 'men act' in these spaces, places and contexts, and they hold their emotions tightly in their bodies and that holding is sacred to the coming community. Contrary to being universal and uncontestable, men's (and women's) emotional experiences are differentiated through complex and coercive associations with spaces and places. This is where Deleuze's affect finds its power. But what is on the other side of that power?

Finding Power in Surrender

Benjamin and I talk at length about the birth of his second daughter and the slow, painful dissolution of his marriage. As we come to the end of our time with the audio-recorder switched on I ask him if he remembers any stories that exemplify his idea of himself as a father.

"Are there things you can remember," I ask, "stories you can tell me that highlight and accentuate an awareness of yourself and your fathering?"

What he comes up with is a story of face-to-face, memory-pumped emotion and a beautifully un-choreographed spatiality as part of the action.

"You know the first memory that comes up is an unpleasant one but it is one that I've grown a lot of strength from because of what I learned years afterwards. Flash-forward to my [youngest daughter] being – at this point we've moved into [a new] house that we've bought which is … in the same neighbourhood. It is a quarter of a mile from our other house. [My youngest] couldn't have been more than three, but I was busy and overwhelmed and, you know, the harried parent of two young kids. The too-busy-career syndrome. Too much too much too much. Plus the tension I was having with my wife and my own life frustrations. My youngest came padding down the hall one day and I remember just being on the floor and looking through a pile of papers trying to find something and she said Daddy, 'can we dah dee dah dee dah,' and I just snapped and I jumped up and said, 'NO, I've got to finish what I am doing,' and she, she jumped back physically and had this horrified look on her face. Phew, and it slayed me. I was horrified by what I had done and I jumped up. She went running back to her bedroom and I went running after her and I scooped her up and I was crying and I held her and just apologized and said I was terribly wrong and told her I loved her and, you know, I was so, so sorry and held her and whatever it was she wanted to do at the time we did. My paperwork had to wait. For the longest time I carried so much regret and guilt about that and, and the context of that … and one episode like that is not going to damage a child especially when you go in and apologize and give her love and attention and never repeat that sort of wounding. Then every other time you apologize – wherever/whenever you transgress – and you are always *present* to that, that really is what will stay in. And that …"

A long pause and a heartfelt sigh.

"… as I look at my relationship with my youngest now, you know I trust that. She really does trust me. So there is always that part of me where I, I wasn't in control and it came out. I remember exploding at my oldest on one occasion. She really could get my buttons, she was really good. And those, those stand out for me. They were formative in maybe not the right ways but in great learning ways."

"Formative events that highlighted your parenting," I emphasize. We both smile hugely.

"You know," Benjamin is thinking about the times he remembers failing as a father and how these are not necessarily failures, "far from viewing myself as a perfect father – and I could probably pick up my mistakes more than anybody else and wish that they weren't there … and yet …" Benjamin pauses as if reflecting on some distant potential.

"You know my situation is very strange because I am not really involved in my kids' lives anymore." Benjamin and his ex-wife are now separated three years. With the divorce, she got custody of their two daughters and although Benjamin sees them, of late they are showing more and more reticence to share their lives with him. The oldest, in particular, wants little to do with her dad. Benjamin shares an example.

"Like both of them, when they spotted me at the Middle-School concert and each of them were in different bands, both of them said to me later, 'What were you doing there? I don't want you there.' And to hear that and take it and not be hurt by it and acknowledge that I am not really in their lives. And yet …"

Again, the move towards potential.

"So that automatically makes it weird and different, like I never thought I'd be here. How could this have happened? Somebody who has been involved in their kids lives, that couldn't be more important. And yet, wow, okay, and knowing that when I am with at least the youngest of my daughters, I hope. And if the other would give me a chance I'd do the same with her. I really, … I listen to her."

I say: "I am sure you do."

"So, you know, it is like my situation is far from perfect. My behaviour is far from perfect and I can think of the little bit I spend with my daughters, they get a lot of validation. And I guess I am going to find out in the years to come if I get their love."

I respond: "And you know what they said to you when you were at the bands does not detract from who you are as a father."

"Right. I know that." Benjamin looks at me with some assurance in his eyes.

"And although your context is different from other fathers I am talking to, it is not so different in terms of the emotions of fathering," I say.

"Yeah, you know I, I, I want them to be prepared to make the right, to make the best choices. You know my therapist has a great saying – it is not his – but it goes: 'We do not raise children. We raise adults.' So, you know, I want them to believe in themselves and behave from that place, and really experience their greatness."

"Do you see that as something that is more important for a dad to do than a mother?" I ask.

"With daughters, maybe. With sons ..." Benjamin pauses here and changes direction in his thinking. "No ... each person is different, we're stereotyping. 'Cause I think in some ways I am way more nurturing than my ex-wife and yet that is typically the feminine, that the woman carries. So she is going to be there for them as a mother in ways that I can never be there for them as a father, yet I am glad she is there for them. I think, I think there is an element of the masculine in her in terms of discipline and making choices ... something I never had any role model on. So I am doing this by the seat of my pants but it feels right."

"But to suggest that," I retort, "and I am not saying you are suggesting this, but sometimes there is a stereotype that suggests that fathers have difficulty being nurturing and that is the woman's job. And these interviews I am doing suggests that is not the case. There is a nurturing that is fathering energy that is equally compassionate and as richly based in emotions as anything that I've come across that relates to mothers."

"Yeah," returns Benjamin, "I think it is important for both parents to do that. You know the greatest gift would be to know that my daughters could turn to me or their mother with any deep problem that they have. Time will probably tell. Because they will have problems. Because we can't avoid them, right?" He laughs and it is a joyous laugh in the face of knowing, and yet not knowing.

A More Productive Politics

I move from the interview with Benjamin with mixed emotions. These interviews are to a large degree cathartic, and they are also a spun coin of illness and remedy. In the face of the hegemonic patriarchal domination of striated space within which the idea of fatherhood resides, more materially realistic notions of disorder are welcome as the keys with which to unlock difference. Put another way, the multiplicities of masculinities become irreducible to identity as their dimensionality constantly shifts. As a result of such movements, fathering becomes de-centred and enters into new relations with itself and others. It is no longer the same because the new embodiments that comprise it encounter one another from new angles. Rhizomic roots continually decompose the unity of masculinity into a thousand microsites where forms of contestation, cooperation, and indifference occur. These more transient geographies at minimum suggest the flows, resistances, consent and dissent that typify material realities of embodiment and experience.They comprise the love and the fear and the anger and the joy and the cuddles and the curses and the shouts and the silences.

Silences

Back to my dinner with Ed. I am interested in where Ed is going with the notion of silences as a potential way of breaking down awkward categories. He is busy worrying a portion of steak on his plate so I take the opportunity to chime in with one of the themes that defines the *Awkward Spaces* project.

"I think one of the problems is that the whole notion of dualism is pervasive in our society so men looking after children define themselves as not-women and stereotypes are trumped over the complexities and subtleties of being a father."

Ed sits back in his chair after carefully wrestling the piece of steak from his plate and popping it into his mouth. He chews thoughtfully.

"You know I've thought about that before and, and there's a … dialectic that … speaks … of understanding the world, not in terms of the way the world is …"

Ed's voice trails off in contemplation to return forcefully.

"There is a huge difference … and that is the way we understand one another. We learn about one another because it is so subtle and complex and profound, it is like choosing words, you know? I mean every single word we use is a category but what else have we got, you know? But we do have other things and with people it is alternative ways of communicating that do not involve categories and those tend to be those silences and eye-contact and all those things, and body contact. All the tactile, all the other senses and stuff where the brain is kind of turned off and it is just a direct transmittal, you know?" Ed grunts in a contemplative way. "But that in itself is not enough to get on in the world."

PART II
Closing In

Chapter 4
Cinematic Landscapes and Leaking Bodies

With this chapter I take a break from ethnographies and turn to cinematic representations (which I treat from a non-representational perspective). There are many cinematic renderings of fathers and their spaces. This chapter looks at two (*American Beauty,* 1999 and *There Will be Blood*, 2007) under the motifs of bodies and landscapes and Chapter 7 looks at two more (*A Perfect World*, 1991 and *Paris, Texas,* 1983) under the motif of journeys. All four movies are award winners of varying kinds (Cannes' Golden Globe, Oscar) and they are chosen in part for their wide critical appeal. Each tell partial stories of fathering that I call thumbnail sketches. The first two movies embolden my appetite for saying something about the capitalist-based embodiment of fatherhood and its quirky relations to US landscape ideals and imperialism. This then takes me back to ethnographies in Chapters 5 and 6 where I say something about new forms (embodiments) of fathering. These forms suggest beginning journeys (lines of flight, if you will) that are then taken up pictorially in Chapter 7 with the second pair of movies.

But why thumbnail sketches? Why break up the pointed actuality of the ethnographies with cinematic representations?

Thumbnail Sketches

bell hooks (1996, 1) notes that movies are like magic because they change things; they take the real and make it into something else right before our eyes. She argues that audiences do not get a dose of reality from movies, they get precisely the opposite; they get a recreated, re-imagined reality.

Movies may look like reality and they may feel like reality as my emotions are engaged, churned and chewed by what I see on the screen, but they are something more. As I am drawn into characters and their stories, it is not a mirroring of reality that moves me but something entirely different and that is what makes movies so powerful. It is not just the images that are re-imagined. As Henri Lefebvre (1991, 10) puts it: "the spectator is uprooted from his everyday life by an everyday world other than his own … this explains the momentary success … films enjoy." I want to escape the pressures of everyday life and I want entertainment; I also want to learn and I may want to gain cultural capital from engaging with the latest and most acclaimed spectacles. From this perspective, viewing movies is in part about change and transformation that is illusive and difficult to tie down.

And connecting with movies is also about sameness. It is about elaborating and blowing apart the fixity of identities and it is about endorsing my world-view, for otherwise I could not relate to the characters and their stories. hooks (1996, 2) points to movie viewing as simultaneously about cultural reification and resistance:

> In this age of mixing and hybridity, popular culture, particularly the world of movies, constitutes a new frontier providing a sense of movement, of pulling away from the familiar and journeying into and beyond the world of the other … Movies remain the perfect vehicle for the introduction of certain ritual rites of passage that come to stand for the quintessential experience of border crossing for everyone who wants to take a look at difference and the different without having to experientially engage 'the other'.

The movie images and their narratives are small renderings – thumbnail sketches – that I blow up (blow apart) with a reading of larger significances, larger cultural connections. And so, with this chapter I bring together affect and bodies and landscapes and consumption by presenting a reading that highlights differing worlds of fathering. In what follows, I use *American Beauty* and *There Will Be Blood* to provide a sense of movement and, in particular, a sense of pulling away from familiar fathering frames. But there are also important elements of familiarity in the movies I choose to discuss. There is familiarity in some of the depicted landscapes, and in the way characters relate to each other and to the places they inhabit and travel through. More importantly, I argue that these movies provide a window (sometimes only a glance) at the possibility of fathers becoming other.

I begin with the face of the father and aspects of his body.

The Face of the Dead Father

Julie Kristeva (1986, 296), whose work on the abject body and the co-mingling of bodily fluids features in this chapter, writes that "[o]ur present age is one of exile. How can one avoid sinking into the mire of common sense, if not by becoming a stranger to one's own country, language, sex and identity? Exile is already in itself a form of *dissidence* ... a way of surviving in the face of the *dead father*." The face of the dead father is – like the face of capitalism – an imagined chimera, a constantly changing spatial frame. And, as I note earlier, for Gibson-Graham (1996, 2006) the imaginary of capitalism (and patriarchy) as a monolithic leviathan to be resisted and revolted against does not serve well a geography of hope and care. Similarly, Meagan Morris (1992) argues famously that a psychoanalytic rendering of the phallic cityscape with its high-rise towers and monumental obelisks represent the Dead White Father in facile ways. She points out that the tower must be something other than a phallic symbol because it is easily exposed as an unseemly, laughable penile display. Rather, Morris argues, these phallic jokes serve to mask the brutal, viscous, complex and contradictory power of corporate capital by reducing urban spaces to a singular, monolithic psychosexual edifice.

Steve Pile (1996) moves Morris' argument further by arguing that the power of capital is not a phallus but an ever changing face, a façade, a spectacle. Allegorically, Pile argues, the substitution of the body of the White Man by a disturbingly androgynous face may be read as the substitution of the brutal, penetrating body politic of urbanized capital by "the cityscape as a collection of postcard scenes" (1996, 223). The skylines of Manhattan and Los Angeles, and other global cities, become the perpetually acceptable face of capitalism.

Mid-way through the chapter, I return to the face of capitalism as an imaginary of fathering that journeys beside a stoic patriarchy in the service of a larger spatial frame (a monolithic global capitalism), but I begin by looking at the ways fathers' bodies are constructed and contrived.

After establishing the context of exile in the face of the dead father, Kristeva (1986, 298) goes on to suggest that "[a] woman is trapped within the frontiers of her body and even her species, and consequently always feels exiled both by the general clichés that make up a common consensus and by the very powers of generalization intrinsic to language." This chapter is a visual rendering of the exile of men as fathers through their bodies. I try to construct something that is more hopeful than Kristeva, Morris or Pile suggest; something that embraces the strangeness of fathering identities while at the same time moving beyond the prison cell of men's bodies and their positionings.

Embodying Fatherhood

If I accept Pile's (1996) suggestion that capitalism today is embodied by an androgynous face rather than the body of the Dead White Father then interesting questions arise regarding the ways fathering bodies are now represented. Gibson-Graham (1996, 101–3) argue that modernity is grounded in Man's body "constituted as an organism structured by a life force that produces order from within." Man's body "became ... the modern *episteme,* setting unspoken rules of discursive practice that invisibly unified and constrained the multifarious and divergent discourses of the physical, life, and social science." And so, from this, modern economics is grounded in Man's body: "... [it is] an economy that is organically interconnected, hierarchically organized and engaged in a process of self-regulating reproduction." The modern capitalist economy is akin to a particular kind of representation of the male body in that it maintains itself by subsuming or displacing its exterior, and by producing integration and wholeness as an effect. It may be argued that this embodied connection to the roots of modernity is also connected – through Rousseau and the other Enlightenment thinkers – to assumptions about men in relation to families. What, precisely, might these relations look like?

At one level, men's bodies are the means through which things are built and fixed and through which harm is done. Their bodies perpetrate violence and are victimized by violence. Men's bodies do not produce children directly, but rather they produce the seed through which life is created, through which it germinates and grows. From that point, Rousseauvian thinking suggests, men are not necessarily connected bodily to the life of an infant. That men are supposedly less connected to infants (and their own bodies) and are, as a consequence, less connected to something called 'nature' moves problematic assumptions about men's ties to families that are rehearsed in the arguments of Popenoe (1996) and Blankenhorn (1995) from Chapter 2. Suggested from those readings is the Rousseauvian notion that men do not naturally connect to children and families and must be enticed to maintain a commitment. That Popenoe (1996, 169) expresses the view that men's bodies are not biologically attuned to infant care – they do not menstruate, give birth and express milk; they are not curved for balancing babies nor soft for cuddling – presupposes a different purpose. One problem with this way of thinking is that it does not give sufficient credit to what fathers do, and how their emotional work inspires action.

A characterization of the male body as strong, hard and impervious ties with notions of building, providing and defending in opposition to notions of nurturing, caring and connecting emotionally. And this comes across in movie representations of fathering. Robin Williams' character in *Mrs. Doubtfire* (1993), for example, is unable to access his estranged children unless he dresses in drag and masquerades in the body foam of an elderly Scottish nanny. In the guise of Mrs. Doubtfire, Daniel Hillard is employed by his wife – busy with her high-earning executive career – to look after their children. Much comedy is made of Hillard's body foam: it is huge and frumpy under a balloon-like dress; enormous foam breasts catch fire over the stove, the act of peeing is recast as an acrobatic performance, the grey nanny wig has a grotesque life of its own. The irony of the film is that it is only from within the body suite that Hillard is able to act as a responsible, authoritative parent figure who – like Mary Poppins – brings order to a chaotic household and wins his wife's respect and gratitude.

If, as suggested in Chapter 2, the work of Popenoe and Blankenhorn is exemplary of academic writing against fatherless families, dead-beat dads and bad-dad syndromes that emerged empirically from the 1980s onwards then it is worthwhile noting the contemporaneous ground-swell reaction that worked against the images of Mrs. Doubtfire (even the name suggests struggle) with a characterization of fathers' roles as excessively connected to their children. These 1990s images of fathers and infants are hugely embodied (e.g. Figure 4.1).

At one level, we have men taking on traditional women's roles and forms, and at another level a salvo comes from the embodied strength of father figures. After Spencer Rowell's iconic picture of a half-naked man (who is muscular with a face half in shadow) holding a baby (who is looking up into the father's face) appeared in 1991, there were a plethora of media images of male sports heroes and Hollywood fathers holding babies. Of course, there is another reading of these images that relates to the colonization of children's lives by men/fathers, men's bodies and capitalism at a time when men were chastised as absent from their children's lives.

But the issue I want to raise here is that unlike women's bodies that are often characterized by leakage (through cycles of menstruation), men's embodied

Figure 4.1 Spencer Rowell's iconic depiction of man and infant
Published with kind permission from Spencer Rowell

representations from the 1990s onwards are still structured in precisely the ways Gibson-Graham (1996, 101) articulate: as a life force that is intact and that produces order. In a moment, I'll use *American Beauty* and *There Will be Blood* to offer another set of possibilities, but first it is perhaps useful to elaborate more fully the cultural battlefield that comprises bodies.

Bodies as Battlefield

Using a curiously masculinist metaphor, Kirsten Simonsen (2000, 7) argues that bodies are a "cultural battlefield." Using similar metaphors, Felicity Callard (1998,

387) notes that within contemporary "fights over theory," bodies are "pursued, wanted, tugged, grasped, and torn apart ... a nearly endless resource with which one may produce compelling theory, harvest tropes, anchor one's desire to protect and articulate 'difference'." The reason for this interest is complicated and compelling; bodies, notes Ruth Butler (1999, 239) are "an active and reactive entity which is not just part of us, but who we are."

Simonsen (2000) sketches three arenas through which studies of the body entered geography. First, the body as "the geography closest in" is an obvious endeavour. Second is a need to understand the body as it is used to "other." Third, a focus on bodies is an attempt to transcend the Cartesian mind/body split. Relational conceptualizations about "the geography closest in," argues Simonsen, jettisons any ideas of scale to focus on how bodies are made and used, the ways they labour, and how power is inscribed on and resisted by bodies. This perspective aligns with Foucault's (1977) claim that socio-political structures construct particular kinds of bodies with specific needs and wants. From this perspective, bodies are primary objects of inscription for societal values and mores. A broadening of "the geography closest in" metaphor uses corporeality to encompass a material base for understanding the social construction of fathers. The construction of "other bodies" is suggested by Mrs. Doubtfire and the Spencer Rowell poster; they acknowledge not only the differences but also the power relations in embodiments. The body is central to how hegemonic discourses designate certain groups as "other" and how fathers are placed in a variety of categories (e.g. able-bodied, obese, bespectacled, minority, muscular). If men's bodies are sources of strength, they are also sources of insecurity and feelings of inadequacy. They are symbolic purveyors of competence and incompetence, sites through which intimacy is experienced or thwarted, and instruments through which difficult emotions are communicated, concealed and contained (Gadd 2003, 351).

An understanding of "the geographies closest in" and "the construction of other bodies" clearly overlap and it is not necessarily useful to try to set them apart. Simonsen (2000) notes that both perspectives come together in a critique of the hegemonic dualisms that permeate Western culture. And so, a focus on bodies is also an attempt to transcend the Cartesian mind/body split and the ways it slips into other dualisms such as culture/nature, public/private, man/woman, father/mother, constructivism/essentialism, adult/child and so forth. Gibson-Graham (1996, 101) note the ways that the story of Man and his body is a colonizing regime through which a foundational set of binaries are able to capture and subsume others. Of course, the mind/body dualism permeates a large swathe of Western philosophy since Plato's belief that the mind dominated matter and that the acquisition of knowledge required that the body be disciplined by and subjected to the mind. But it was from a small room in a 17th century provincial French town that René Descartes – in a lonely chamber, longing for release from the body's encumbrance (Zita 1998, 166) – developed a foundation for modern scientific knowledge that required the separation of mental discipline from the seemingly irrational and certainly unruly passions of the body. He inscribed on the body a modernist trope that structures corporeality as a mechanical substance reducible to itself. Descartes expanded knowledge, step by step, to admit the existence of God (as the first cause) and the reality of the physical world, which he held to be mechanistic and entirely divorced from the mind. This is almost a complete dualism because it requires a separation of the material body from the inner self, and creates an ontological gulf that is bridgeable only by divine intervention.

Even the focus on emotions over reason from Enlightenment thinkers such as Hume and practitioners such as Owen (see Chapter 2) was not sufficient to dispel the Cartesian project. With the dominance of this project through modernity, it is possible to understand what Gibson-Graham (1996, 101) call a "bizarre dance of dominance and submission through which Man addresses the economy." To paraphrase Gibson-Graham, if Man-as-father is positioned as the first term in the binary, he is the patriarch, the master of his and everyone else's destiny through logic and reason; but when Man-as-father is positioned as the second term, he bows to the patriarch as to his god. Each positioning is reinforced by an infinity of representations in the same way that fractal patterns appear again and again up and down some prescribed hierarchy: for example, man (mind)/father (body) or father (god)/man (humanity), *ad infinitum*.[1]

Of course, this is only one way to fantasize the body's matter. While the Cartesian project gives mind, masculinity, rationality and sameness priority over the body, femininity, irrationality and otherness, corporeality and bodies are nonetheless always present although their role is complex. Simonsen (2000) notes that critical discussions in geography, and most of the other social sciences, focus on embodiments and discourses about the body with very little attention paid to the material body. Rather, priority is given to a deconstructive project that attempts to dismantle the Cartesian mind/body split or destabilize hegemonic notions of the body and dominant discourses on embodied identities. Simonsen notes that there is a certain amount of ambivalence towards material bodies and individuals' everyday interactions with their bodies. This ambivalence is noted also by Jeff Hearn (1996, 212), who argues that "to assume *a priori* that masculinity/masculinities exist is to reify the social construction of sex and gender, so that the typical dimorphism is assumed to be natural." And like Gibson-Graham, he sees how this assumption leads to other dimorphic structures that then reproduce a "heterosexualizing of social arrangements." Instead, like Simonsen, he argues for exploring the material basis of masculinity by questioning how the 'quality' of masculinity relates to what men do in work, in fathering, in sexuality, in violence and elsewhere. This seems to me like a very good idea.

A move to materiality suggests that (at least to date) tracing politics onto a body space is effected by how the body is seen and recognized, but not by how the body feels (Pile 1996). A non-material perspective is limited by an enduring

---

1 In Chapter 6 I elaborate more fully the context of naming and repetition from a Butlerian and Deleuzian perspective. See Lukinbeal and Aitken (1998) for an elaboration of fractal representations and the persistence of patriarchy in mainstream movies.

focus on representations and tracings rather than mappings (to use the Deleuzian distinction I raised in the introduction). Through the 1990s geographers and other social scientists with interests in the power of images, became suspicious of the postmodern penchant that focused on power through the camera's eye. While Baudrillard (1988) famously quipped that there was a planned depthlessness in contemporary culture that was no more than dancing lightly on the celluloid skin of the film frame, other commentators were increasingly sceptical of research stances that elided the material conditions of lived experiences. Mapping politics onto body space is not just about representations of the body, but how those representations affect us. Taking note of how the body is affected requires perspectives that are non-representational; these are an intensely political derivation from Massumi's (2002, 495) axiom that the "skin is faster than the word."

If the contemporary battlefield acculturates us to the notion of hard, impervious male bodies that can be relied upon, then what happens when those bodies are represented as squishy, leaky or constantly in motion? In what follows I suggest that *American Beauty* reworks some notions of fathering in quirky ways by setting up some familiar American tropes of middle-class suburban landscapes and then re-positioning a father's body in unexpected ways.

The Skin is Faster than the Word

British stage director Sam Mendes plays with the notion of fathers' bodies and their leakages in his film-directing debut, the award winning dark comedy *American Beauty* (1999).

Heralded as a move back to intelligent film-making, the movie won five Oscars, including Best Picture.

Kevin Spacey plays Lester Burnham, a cynical 42-year-old father living in a middle-class American suburb. Lester narrates the movie – "my stupid little life" – from the grave, beginning with an oblique aerial shot of his neighbourhood and then seemingly floating/flying down to the front of his house, a typical white suburban home with a red front door, the colour of the "American Beauty" roses that are grouped strategically behind a white picket fence:

> LESTER (in voice over): My name is Lester Burnham. This is my neighborhood. This is my street. This… is my life. I'm 42 years old. In less than a year, I'll be dead. Of course, I don't know that yet.
> (pause)
> And in a way, I'm dead already.

Through Lester, the movie explores a variety of themes including paternal love, freedom, sexuality, the search for beauty and the possibility of happiness in relationships. Set in anywhere USA (the movie was filmed largely in Torrance, California), the neighbourhood and family scenes depict beautifully how the banality of suburban life can affect parents and teenage children. Edward Guthman (2000, C-17) of the *San Francisco Chronicle* called it "a dazzling take of loneliness, desire and the hollowness of conformity." Ultimately, the movie is a humorous, existential romp through mundane middle-class suburbia that increasingly moves, with grinding bathos, to an appreciation of the beauty behind all things.

We are propelled to an appreciation of beauty through Ricky Fitts (Wes Bentley), the teenage son of the Burnham's new neighbours. A focus for beauty is Ricky's video-recording of a plastic bag dancing in a spiral of wind and leaves.[2]

From the dancing bag, Ricky speaks to Lester's teenage daughter Jane (Thora Birch) about the beauty he sees in unexpected places, like in the face of a dead homeless person.

> RICKY: Have you ever known anybody who died?
> JANE: No. Have you?
> RICKY: No, but I did see this homeless woman who froze to death once. Just laying there on the sidewalk. She looked really sad. I got that homeless woman on video.
> JANE: Why would you film that?
> RICKY: Because it was amazing.
> JANE: What was amazing about it?
> RICKY: When you see something like that, it's like God is looking right at you, just for a second. And if you're careful, you can look right back.
> JANE: And what do you see?
> RICKY: Beauty.

Ricky's obsession with video-recording replays the well-worn postmodern theme of 1980s movies (e.g. *Sex, Lies and Videotape* (1989); *Videodrome* (1983); *Scanners* (1981)), but in *American Beauty* the focus is more on highlighting and appreciating beauty in the everyday rather than some penchant for created simulacra taking over our lives.

Ultimately, *American Beauty* is about beauty, but there is nonetheless a nod to the military/industrial complex and the power of surveillance. Ricky's dad is an ex-Marine colonel who strictly (and violently) controls his son. In his search for beauty, Ricky's video-recording is also a form of surveillance on his neighbours, and, in particular, Jane. Ricky is caught in a bind: That Jane is eventually seduced by Ricky does not detract from the fact that he transforms the beauty of ordinary experiences by distancing them on film. This is his escape; the freedom of the viewfinder (Smith 2002). The point is that Ricky understands the limits of the medium and when he gets to look at beauty in Lester's death, his camera hangs listless from his wrist.

---

2 Alan Ball originally wrote *American Beauty* for the stage, inspired by a paper bag he saw floating in the wind near the World Trade Center Plaza in New York.

Neither Lester nor his wife Carolyn (Annette Bening) are happy in their work or their marriage and each blames the other for their problems. Their constant bickering across the dinner table costs both of them the respect of Jane. Similar to *Mrs Doubtfire,* the mother in the family relationship is the main 'breadwinner' in the family, although in *American Beauty*, the Annette Bening character is a burnt out real-estate broker who seeks advance (or, at least, an emotional jump start) through a torrid affair with the local real-estate 'king', Buddy Kane (Peter Gallagher). This is her escape.

> CAROLYN: Yes! Oh, God! I love it!
> BUDDY: You like getting nailed by the king?
> CAROLYN: Oh yes! I love it! Fuck me, your majesty!

Meanwhile and, ostensibly, on the same day that his wife's peccadillo begins, Lester quits his advertising job and – after securing through blackmail a hefty severance package – buys the car he always wanted: a red 1970 Firebird Pontiac Transam. His new job flipping burgers provides him the "least possible amount of responsibility." This is his escape.

Moving towards the bathos of the movie, Lester ultimately engages full responsibility for his life through love and joy:

> LESTER (in voice over): Guess I could be pretty pissed off about what happened to me. But it's hard to stay mad when there's so much beauty in the world. Sometimes I feel like I'm seeing it all at once, and it's too much, my heart fills up like a balloon that's about to burst. And then I remember to relax, and stop trying to hold on to it, and then it flows through me like rain and I can't feel anything but gratitude for every single moment of my stupid little life. You have no idea what I'm talking about, I'm sure. But don't worry – you will, someday.

An important part of the freedom that Lester engages is represented in a journey through his body. This appreciative father is leaky and open. For example, one of the first shots of Lester highlights masturbation as a banal and joyless act.

The masturbation scene is bookended with the scene of Lester's murder at the end of the movie. That too focuses on a leaky body, but this time the moment is far from banal. After hearing a gun-shot, Ricky and Jane come downstairs to find Lester in the kitchen. He had been looking at a picture of his family in happier days and reflecting on the beauty of his life. As he gazes at the photograph, which is framed on the right, a gun barrel appears to the left, just behind his head. The murderer is Ricky's dad, Colonel Frank Fitts. Earlier that evening Colonel Fitts had thrown Ricky out of their house in a homophobic rage, believing that Ricky was having an affair with Lester. In the pouring rain, Colonel Fitts had approached Lester in his garage and tried to seduce him, revealing his own latent homosexuality. When Lester backs off telling him he has the wrong idea, the Colonel fits leaves in confusion and shame, to return later with the gun.

Figure 4.2 Lester leaks one last time

Lester's brains are splattered over the kitchen wall and his blood pools on the table and drips to the floor. Colonel Fitts leaves. Moments later, Ricky and Jane come into the kitchen. Ricky bends down to get a closer look at Lester's face. Jane follows him, in shock. Ricky kneels, gazing at Lester's face and then he smiles, ever so slightly. Lester is looking back at him, his eyes are lifeless but he's smiling the same slight smile. Ricky whispers "wow" as he connects with beauty.

The scenes between these two moments of body fluids spilling/spurting out over everyday household objects – between the masturbation scene and the murder scene – are to a large degree about Lester trying to find himself through his body. He pulls free-weights out into the centre of his garage and starts to work out. This action may be interpreted, along with purchasing the muscle-car, as a part of his mid-life crisis or a vain (as it turns out, not so vain) attempt to attract and seduce Angela (Mena Suvari), Jane's nubile school friend. Along with pumping iron in his garage, Lester starts jogging. While running one day, he joins his two gay neighbours, Jim and Jim:

> JIM #1: Lester, I didn't know you ran.
> LESTER: Well, I just started.
> JIM #2: Good for you.
> LESTER: I figured you guys might be able to give me some pointers. I need to shape up. Fast.
> JIM #1: Well, are you just looking to lose weight, or do you want to have increased strength and flexibility as well?
> LESTER: I want to look good naked.

The portrayal of the male body as strong, hard and impervious is an important component in problematic fathering representations. Men are hard and closed and predictable, like granite. So what happens when they don't do what they are supposed to do? What happens when they leak?

Leaking and Co-mingling

Abjection, as defined by Julie Kristeva (1982), is derived from something I do not like but that I desire and from which I cannot divorce myself. Desire and repulsion simultaneously guide relations between insides and outsides, self and other, me and you. Like Ricky, I am drawn to the beauty of the abject, but there is something more. The important point about Kristeva's notion of the abject is that it does not respect borders and definitions and, as a consequence, it leads me to a consideration of difference. Self, other, male, female, black, white, father, child become tenuous categories of relation and Cartesian boundaries predicated upon a mind-body split are blown apart. Abjection – desire and revulsion around something that is in actuality part of me – moves beyond the colonization effects of Cartesian binaries. It moves through the binaries that describe the social construction of sexism and racism.[3]

Carol Clover (1989) raises Kristeva's notion of abjection to explain the appeal of slasher movies and why Final Girl is an attractive hero to teenage males.[4] *American Beauty* does something different. It is not a slasher horror film and the violence and leakages, rather, cut across a rigidly defined social community. David Smith (2002) argues that the movie's characters push against a world that

---

3 It also partially explains the psychological basis of sexual fetishes and hate crimes.

4 Carol Clover (1989) makes the interesting argument, from Freud, that movie representations of male openings (such as eviscerations or their equivalent) are akin to castrations, to emasculations. In her classic discussion of slasher horror films, Clover's (1989) concern is how to explain the appeal to a large male audience of a film genre that invariably features a female victim-hero (such as Sigourney Weaver's character in the *Alien* series or Jamie Lee Curtis in the *Halloween* series). Clover characterizes these women as 'Final Girl' and notes that men's ease at engaging with them is an indication of cross-gender identification. But this is too simple an explanation because Final Girl's masculine interests, her sexual reluctance and her apart-ness from other girls construct her gender. Perhaps most important, her unfemininity is marked by her abilities with "the active investigating gaze normally reserved for males" (Clover 1989, 93). With considerable aggression, the Weaver and Curtis characters search for the monster/mother, track it down to its womb-like lair, look at it, and thereby bring it fully into our vision. According to Clover (1989, 94), the association between Final Girl and the monster is a "shared masculinity, materialized in all those phallic symbols – and it is also a shared femininity, materialized in what comes next: the castration, literal or symbolic, of the killer in her hands." Clover points out that this encounter puts Final Girl and the monster on terms that are about something more than sexual repression and fear of castration, they are also about a loathed shared identity that comes from abjection.

is carefully structured by Sam Mendes as a culturally deterministic system. The characters are perfect creatures of deterministic social relations (mother/father/daughter) prescribed by a series of individualistic capitalist norms. The Burnham family members, in particular, seek to mitigate their circumstances through drives, values and impulses acquired from the cultural – consumer/entertainment – complexes from which they seek escape. This too is about bodies. While Lester pumps weights, Carolyn fucks the King of real-estate and Jane saves up her baby-sitting money for a boob job. The leakages from Lester's body are precisely about the possibility of spillage out of that deterministic system. Taken as a whole, the film, Smith argues, is a meditation on the dislocation between the narrative quests of its characters and the meaning that, in a few cases (Lester, Ricky, Jane), is pushed onto a different path. The burning question of the movie is that if escape and freedom do not reside in the culturally determined places, where do they live? As Smith (2002) questions, if subject meaning is not found through emancipatory projects, then where is it to be found? If freedom does not consist in doing what I want, then what is it? For Smith, emancipation comes through Ricky's engagement with beauty in the dancing bag, or in Jane's sad face, or in Lester's death. It brings with it – for the characters who get it (Lester, Ricky, maybe Jane) a sense of meaning that is unquestionable, ontologically ultimate, absolute and, above all, spiritual. In the moments when Ricky feels that God is looking at him, beauty is what he sees when he looks back.

Façades and Expressions

Beauty exists in a number of different facial forms.

In suggesting that the face of capitalism is now a picture postcard of civic beauty resembling the skyline of Manhattan, Los Angeles, Singapore, Beijing or Sydney, Steve Pile (1996) suggests a move from the identifiable white body of the patriarch to an androgynous façade, which hides the brutality and cruelty of the modern capitalist economy. It is like looking at the beauty of Fritz Lang's futuristic Manhattan skyline in *Metropolis* (1926) or Ridley Scott's Los Angeles in *Blade Runner* (1982). From these skylines is structured – framed spatially in an obvious hierarchy of buildings and towers – the sorrows, darkness and oppression of working people under the giants of capitalism (Frederson and Tyrell respectively for Lang and Scott). Like Lester's dying face but with a different affect, the skyline/façade embodies content and expression as material reciprocations "defined only by their mutual solidarity" (Deleuze and Guattari 1987, 45). The steadfast and self-assured faces of Frederson and Tyrell are simultaneously the faces of the patriarch/the creator/the scientist the corporate magnate/the hierarchy of the tower. It is a visage that is in control of all it surveys.[5] There is, as Semetsky (2004, 316–

5 And in a superbly iconic Oedipal moment towards the end of *Bladerunner*, Tyrell's face of smug assurance is transformed to terror when he realizes that his creation/son

7) suggests, "an intensive multiplicity" at work here; the face/façade "becomes effective and expressive as long as the form of expression is *not separated* from but is supplemented by the form of the content." As expressions, faces/façades are "inserted into or intervene in contents, not to represent them but to anticipate them or move them back, slow them down or speed them up, separate or combine them, delimit them in a different way" (Deleuze and Guattari 1987, 86). The question of faces/façades is not "what is their meaning?" or "what are they representing?" in a dichotomy of signifier/signified, but rather "what are they accomplishing?" In raising these questions, Giorgio Curti (2008) points out that accomplishment is about material action, it is to figure whether the face/façade is good – does it cause or permit the spectator to think and act simultaneously beside differentiating assemblages of filmic content and expression (e.g. Lester's beatific face), or is it bad – does it inhibit spectators, limit them, neglect differences of thought and action for a return to sameness and redundancy (e.g. Tyrell's smugness and his corporate tower)?

Deleuze argues that most often it is the face or its equivalent (the clock, the city-skyline) that gathers and expresses affect in these kinds of complex, elusive ways.[6] According to Massumi (1992, 152), content is overpowered and it is expression that does the overpowering. Of course, the face of the father is not always the beatific face of Lester on the kitchen table, nor is it always the bemused and then terrified face of Tyrell as he is killed by his replicant creation/son. A striation is smoothed and a smoothing is striated in an endless process of expression becoming content and vice versa. It is from this that a crisis of fathering is contrived, as is the solution. Fathering is a remedy to fatherhood, an illness. This staging of expression and content is an important part of narrative. Deleuze (1986, 108) argues that although "the close-up extracts the face (or its equivalent) from all spatio-temporal co-ordinates, it can carry with it its own space-time – a scrap of sky, countryside or background … the affect obtains a space for itself in this way." The content of the face extracts with it its own spatial frame, its own spatio-temporal coordinates.

The face of Daniel Plainview in Paul Thomas Anderson's *There Will Be Blood* (2007) seems to carry its own spatio-temporal coordinates. It is the face of a calculating oil tycoon, a man driven by his ambition who, time and time again, ruthlessly mimics and manipulates people and communities in the building of his empire. Plainview's face exudes all that is repulsive about that kind of determination and, there is also, simultaneously, a look of desperation, an appeal for love and acceptance. The focus of this look is most often upon his adopted son.

('replicant' Roy Batty played by Rutger Hauer) has returned, has outsmarted him, and is now going to kill him.

6 The face, the pure building block of affect, what Deleuze (1986, 103) calls its *hylé*, from the Greek word meaning "matter" or "content."

The Face of Repulsion and Desire

Daniel Day-Lewis spent a year getting into the character of Daniel Plainview for his Oscar-winning performance in *There Will be Blood* (2007). He captured Plainview's gritty voice by listening to recordings of prospectors from the late 19th century; it is a voice that transforms from a crushing, spitting, vicious rasp to a soft-spoken, condescending, magnanimous lilt. Day-Lewis lived with the characters of John Huston's *The Treasure of the Sierra Madre* (1948) to get a sense of how greed distorts perspectives and relationships. He immersed himself in documentation on, and biographies of, Edward Doheny Jr., the oil tycoon upon whom Upton Sinclair based the novel *Oil* (1927) that was the inspiration for Anderson's movie. Day-Lewis created a focused, self-willed character for Plainview and carried that persona both on and off the set throughout the filming. Of particular note is the facial intensity that Day-Lewis creates in his slightly forward leaning, almost surging, figure. It is a face that brings alive the single-minded greed and avarice of Plainview, with a simultaneous hint – a slight and almost imperceptible glimmer – of human compassion and need. In *There Will Be Blood* that compassion and need is positioned in Plainview's relationship with his erstwhile son, H.W. (Dillon Fransier).

There Will Be Blood is touted as a brooding epic that harkens back to vast landscapes of avarice and greed seen in *Giant* (1956) and *Citizen Kane* (1941), but it is also an intimate engagement with bodies and fathering.[7] The first 20 minutes of the film has no dialogue as we follow, in a painful and fragmented way, Plainview's beginnings as a silver prospector and then oilman. The images are of men at toil, muscles straining in dark, humid, telluric spaces. Bones are broken in falls, skulls are crushed by falling machinery. The oil that covers them is dark, viscous, blood-like. The men emerge from narrow holes as if the earth were birthing them. Plainview gains his son as a baby from a miner who dies in one of his oil pits. In taking H.W. as his own, he uses him as an innocent face beside his grizzled determination to add a semblance of family life while he pitches his oil projects to town-folks.

The confining telluric spaces of Plainview's beginnings are contrasted with the barren landscapes of early 20th century California, that dominate the balance of the film. If the beginning of the film symbolizes Plainview's birth from the earth – an earth that breaks his body and moulds it to 'her' wishes – then the balance of the film is about control of the earth and the peoples who inhabit its surface. The face of Plainview is the face of an empire builder, a self-absorbed pioneer who traces a raging capitalist enterprise onto the undeveloped landscapes of California. Plainview's face affectively parallels the façade of the wooden oil derricks, new town buildings, pipelines and, finally, his echoing lonely mansion overlooking the

7 Whereas Sinclair's novel focuses primarily on the son of the oil tycoon and his rebellion against his father by embracing socialism, Anderson's screenplay embraces only the beginning of the novel and the movie focuses mostly on oil landscapes and the ways they intertwine with the avarice and greed of the father.

Pacific Ocean. The face and the façades are contextualized beside barren, forlorn landscapes so untypical of popular representations of California and the American Southwest.

An Embodied, Consumable Empire

In a paper famously presented to the American Historical Association in 1893, Frederick Jackson Turner (1921) bemoaned the loss of the frontier, perceiving it as the cornerstone of American identity. Nature in America was significant not only because its peaks and vast expanses challenged the works of European society, but also because it developed an adventurous, frontier character, prescribed in large part by an indomitable and rugged male individualism. Anderson's Daniel Plainview is emblematic of this emerging discourse and its connection to an imperial capitalist politics. But the landscapes of the American West that Turner alludes to – and Hudson River School artists like Thomas Cole and George Caitlan painted and, latterly, John Ford created cinematically – are very different from the ones portrayed by Anderson in *There Will Be Blood* (see Figure 4.4).

I argue elsewhere that the work of Hudson Valley School landscape painter Albert Bierstadt provides a meta-narrative for Turner's 1893 proclamation by opening up an "Imagined Community" – to use Benedict Anderson's (1991) famous concept of nationhood – that was very specifically tied to landscape (Aitken and Dixon, forthcoming). In short, Bierstadt provides, first, an exotic and dangerous frontier for a people newly mobilized through railways; and, second, a "consumable empire" for patrons back east who wanted to hang it on their walls (Mitchell 1996, 64). Moreover, and importantly for what I want to say here, Bierstadt's work is often credited as a forerunner to the mise-en-scène of the classic Hollywood Western (cf. Mitchell 1996). In the late 19th century, Bierstadt moved to California and began a series of paintings that established his fame as the founder of the "California School of Landscape Painting." With *The Rocky Mountains, Lander's Peak* (1863, Figure 4.3) – the painting that made him famous – and iconic work such as *The Yosemite Valley* (1868) and *Mt. Whitney* (1875), the populace of the Eastern Seaboard (which, at that time, was the main consumer of his work) was treated to a particular rendering of the American West: a spectacle described by some as arousing Wagnerian aesthetics.[8]

8 Through his paintings Bierstadt accomplished four quite startling connections to the imperious manifest destiny of a young, mobile nation seeking a sense of self, a raison d'être if you will. First off, Bierstadt pushed the artistic sentiments of Frederick Church, Thomas Cole and other Hudson River School painters by describing a broad disconnection from the grandeur of ancient European cultural icons – Athens' Pantheon, Paris' Cathedral Notre Dame – replacing them in a developing national psyche with the distinct splendour of America's wild places. The sublime grandeur of European ancient history is replaced with the spectacular geology and meteorology of nature. Secondly, the wild places he painted not only suggested grand natural cathedrals, they also promised fertile and abundant

Figure 4.3 Bierstadt's *The Rocky Mountains, Lander's Peak* (1863)
The Metropolitan Museum of Art, Rogers Fund, 1907 (07.123). Image © The Metropolitan Museum of Art.

The importance of place as an imperious catalyst for meaning creation is elaborated by a number of philosophers, including J.E. Malpas and Edward Casey. If we accept (as most geographers do) that "place is primary to the construction of meaning and society" (Creswell 2004, 32) then it takes only a small extension of this idea to side with J.E. Malpas (1999, 198) when he states that, "the idea of place does not so much bring a certain politics with it, as [it] define[s] the very frame within which the political itself must be located." So the landscape of the American West is defined as a spatial frame from within which identity foments and against which it is difficult to push politically. The American West is more than a geographical region; it is the transcending spectacle that forecloses other political possibilities.

Similarly, in *There Will Be Blood,* Anderson creates a landscape that forecloses upon any transformation other than into a corporate empire. It is not spectacle in the Bierstadtian sense, but it is nonetheless (and perhaps more so), a landscape that

homesteading in pristine, untouched and attainable landscapes for those adventurers willing to challenge wildness. In an important dialogue with American imperialism, the paintings invited settlement and conquest, calling to would-be pioneers with their idealized wide-open spaces and unoccupied, fertile lands. For those less inclined to venture west, the landscapes could be bought and named. Bierstadt's proclivities towards high society propelled him to using large canvases, which excited well-heeled clients with walls to fill. At the height of his fame, Bierstadt went as far as to offer naming opportunities over western mountains for those willing to pay extravagant prices for his paintings.

Figure 4.4 Plainview and H.W. arrive in a barren California landscape

is consumable (Figure 4.4). And, importantly, it is a landscapes that mirrors, and is mirrored in, the face of Plainview. After contriving a suspect deal with an unwitting community, Plainview sets up his drilling operation. Rather than a Bierstadtian spectacle of the American west, the viewer is treated to a dystopian and barren landscape upon which the new wood of the oil derricks provide welcome relief.

Although Bierstadt's painting connects to his imperial and propagandist leanings, it also connects to a later filmic landscape. This is a point I borrow judiciously from Lee Clark Mitchell (1996, 66) who, in his wonderful book *Westerns: Making the Man in Fiction and Film*, argues that, "[p]art of what critics have persistently thought of as a 'problem' of Bierstadt derives from his socially transgressive, metaphysically transcendental yearning – the unappeased craving." In this sense Bierstadt's work is both illness and remedy for the imperial subject. And so there is another way to look at the faces/façades of filmic landscapes – the remedy, a re-territorialization that is not imperial but is connected to important ways that particular emotions are mapped. And for this understanding, as with Bierstadt, I want to go back to pre-filmic constructions of images and maps.

In Plain View

It is appropriate not only to think about structures of desire and repulsion suggested by movies such as *There Will be Blood* and *American Beauty* but also about structures of fantasy and their relations to identity and landscape. Linda Williams (1991, 11) points out that fantasies "are not … wish-fulfilling linear narratives of mastery and control leading to closure and the attainment of desire. They are marked, rather, by the prolongation of desire, and by the lack of fixed position with respect to the objects and events fantasized." Fantasy is a place where conscious and unconscious – interior and exterior and part and whole – meet. It is a place resembling what Kaja Silverman (1992) refers to as a mobius strip reflecting relations between interiority and exteriority, where each interchanges with the other imperceptibly as one move along the strip. Plainview's face and body are part of the consumable empire that Anderson creates in the Californian landscape; in the same way that Lester's face and body are part of the consumable suburban empire that Mendes creates.

In their now classic essay "Fantasy and the Origins of Sexuality," Laplanche and Pontalis (1986) argue that fantasy is the staging of desire, its mis-en-scène rather than it's object.[9] In this sense, it is difficult to untangle fantasies from landscapes. Williams (1991, 11) notes further that Laplanche and Pontalis's (1986) understanding of fantasies is linked to "myths of origins" which moves between two discrepancies: an irrecoverable original experience and the uncertainty of its imaginary revival. In this sense, fantasies moves me 'mobiusly' between interiority and exteriority, consciousness and unconsciousness, bodies and landscapes, the real and the imagined. The important point that Williams raises is that the juncture of this irrecoverable real event and the totally imaginary event has no fixed temporal or spatial existence: it is entirely virtual. Her argument melds well with how contemporary Freudian and object relations theorists, pulling from Melanie Klein and Julia Kristeva, understand 'the origin of the subject' as unfixed, unstable and transitional. Fantasies, then, are about mythic origins and their power derives from contextualizing 'exterior' landscapes as well as the unconscious.

If *There Will Be Blood* is an outward journey of conquest, collusion and single-minded empire building, it is simultaneously an inward journey to sentiment and the work of the father. If the film's dialogue-free beginning feels like a metaphorical birthing that is a violent contest between men's bodies and the earth, it is simultaneously a material birthing of Plainview's son. H.W. is the son of one of Plainview's workers whose skull is crushed by a falling support beam. Plainview takes H.W. on as his son, ostensibly to give him an air of family respectability as he manipulates communities into giving him drilling rights on their land. And so, at one level, H.W. is a tool to further Plainview's ambitions. Throughout H.W.'s boyhood we nonetheless are treated to glimpses of intimacy between father and son. Daniel Day-Lewis's genius as an actor is manifest in the ways that he holds Plainview's face in a visage that is simultaneously about desire

9 In their later book on psychoanalysis, Laplanche and Pontalis (1993, 277) note that people live their lives through spatial relationships that are "the entire complex outcome of a particular organization of personality, of an apprehension of objects that is to some extent or other phantasized."

for, and repulsion from, intimacy. The latter wins out when H.W. is struck deaf in a drilling accident and Plainview loses his lifeline with any kind of intimacy. It severs their relationship (H.W. is sent to a special school in San Francisco) and shakes Plainview's already tenuous hold on sanity. The desire for intimacy with his son becomes an unappeased craving that Plainview fills in other ways. His greed, avarice and murderous ambitions mount over the years as he strives to build a pipeline from his oilfields to the coast.

Another narrative strand runs through *There Will be Blood*, represented by the angelic, boyish face of Eli Sunday (Paul Dano). Eli represents the community against which Plainview struggles. Attempting to gain advantages for himself and his evangelical ministry, Eli sets himself up in opposition to Plainview. At one level, Eli's community represents the old world that Plainview is destroying. That said, Eli's ambition and avarice matches Plainview's stroke for stroke. The two clash at a number of critical junctures in the story. In an epic, and lifelong struggle, each is out to best and humiliate the other.

The penultimate scene of the film takes place at Plainview's coastal mansion and a confrontation with H.W. (who is now adult and, still deaf, works with an interpreter). H.W. wants to take his experience as an oilman to Mexico so that he can establish his own empire separate from his father's. Plainview blows up, curses his son as a bastard and disowns him. The next morning, Plainview is woken from a drunken stupor by his nemesis, Eli, who, in a stunningly cruel scene, he belittles and then kills. "I'm finished," he says as his butler enters and surveys the murder scene. The movie ends with Plainview's face framed in an expression that is simultaneously success and failure.

Embodiment as Illness and Remedy

Illness/success/remedy/failure in *There Will Be Blood* is connected to an emotionally and politically rendered place that is consumable. As with manifest destiny, it is a hunger to move beyond places that confine, and it is part of an imagined community that breaks territorial boundaries, but it also reforms those boundaries. It is about commodification and empire building. It is the face of the father.

It is worth concluding with a comment upon the ways that consumption relates to fantasies and phobias, and how this relation turns on an understanding of inner and outer landscape spaces. Here, perhaps, we gain insight from Lacan.

Jacques Lacan (1978) suggests that with desire I am trying to assert a sense of being, a way to mark my existence. If, in this sense, rather than a lack, desire is conceptualized as a positive source of new beginnings then landscapes/faces/façades get to mark these beginnings and propel them forward. Correspondingly, as a seeming illness emanating from, say, patriarchy, then men/fathers/sons are taught to occupy space in ways that connote strength potency and assertiveness in the oppression of others. The male body is translated into an objective, physical project, subject to the motivation and will of its owner, a view that leads to the achievement-oriented, impervious, self-sufficient masculinity of corporate offices, football fields and commercial pornography.[10] This is the brutal, and often phallic, commodification of the body politic as exemplified with the Spencer Rowell poster. If this is, at times, an illness, I argue that the face/façade also evokes a remedy – the father's freedom, his beauty – in other material ways. And with this I want to argue for a move against spectacle and commodification. A move of this kind comes with affective geographies of belonging and becoming, in the beauty of moral freedom found in Lester's fathering.

Beauty is found in the unlikely and in the abject; in the relations between insides and outsides, self and neighbour, family and stranger. In *American Beauty*, beauty stands apart from the characters' gratuitous commodity quests that are merely illusions of desire, spatially enframed by somebody else (perhaps by a real estate "king"; perhaps by gay neighbours). In the end, suggests Smith (2002), Lester exercises freedom but not freedom as license to do what he wants: seduce Angela. Rather, his freedom precipitates a clarity of moral choice to do the right thing: make Angela a sandwich. For Lester, freedom comes in the form of moral insight. What sort of freedom, asks Smith, is possible from the deterministic system of family and community/neighbourhood presented in *American Beauty*? What sort of liberation is possibly from the trap of family and fatherhood; what sort of freedom is possible from a trap that is what we *are*? Freedom from self, in this sense, is not intentional, but can be experienced, like the beauty in a dancing bag, as an aesthetic that provides an affirming moment. Beauty is found in the necessity of fathering and of partnership within a household, because within these aesthetics – like men's bodies; like insides coming out (a Deleuzian 'fold') – there

10 Commodification is the most important aspect of Lacan's post-Freudian existential sexuality that is yet to be tied into a theme global consumerism. Perhaps this could be done best by evoking Guy Debord (1983, 2000) who argues that the current spectacle *is* the transformation of desire and fantasy into the reality of the commodity occupying the totality of contemporary life:

> The spectacle is the moment when the commodity has attained the *total occupation* of social life. Not only is the relation to the commodity visible but it is all one sees: the world: the world one sees it its world. Modern economic production extends its dictatorship extensively and intensively. In the least industrialized places, its reign is already attested by a few star commodities and by the imperialist domination imposed by regions which are ahead in the development of productivity. In advanced regions, social space is invaded by a continuous superimposition of geological layers of commodities. At this point in the 'second industrial revolution,' alienated consumption becomes for the masses a duty supplementary to alienated production. It is *all* the *sold labour* of a society which globally becomes the *total commodity* for which the cycle must be continued. For this to be done, the total commodity has to return as a fragment to the fragmented individual, absolutely separated from the productive forces operating as a whole. Thus it is here that the specialized science of domination must in turn specialize: it fragments itself into sociology, psychotechnics, cybernetics, semiology, etc. watching over the self-regulation of every level of the process Guy Debord (1983/2000, 42).

is both desire and repulsion. And this is a consequence of a journey. Smith (2002) stages the journey in this way: because Lester quits his job he loosens up; because he loosens up, he pays attention; and because he pays attention, he recovers his world.

But what precisely, is the payoff in embracing desire and repulsion? In the next chapter I explore this through what I am calling the inevitable father and his embodiment in the "essential space of community" (de Certeau et al. 1998, 252).

Chapter 5
The Inevitable Father

> Considering culture as it is practised, not in what is most valued by official representation or economic politics, but in what upholds it and organizes it, three priorities stand out: orality, operations, and the ordinary. All three of them come back to us through the detour of a supposed foreign scene, *popular culture*, which has benefited from numerous studies of oral traditions, practical creativity, and the actions of everyday life. One more step is required to break down this fictive barrier and recognize that in truth it concerns *our culture*, without our being aware of it (de Certeau et al. 1998, 251).

This chapter is entitled 'the inevitable father' as an ironic gesture that points to a 'fictive barrier' in recognizing the importance of communal households and community support in the everyday work of fathering. It is a move away from the isolated fathering practices sketched out in the previous chapter.

Of course, there is nothing natural or inevitable about fathering in the same way as it is problematic to suggest mothering is the more natural form of parenting.[1] The chapter is a challenge to the naturalness of parenting and an attempt to reposition fathering and mothering as a work (a labour, a toil) rather than *the* work of spatial discourses and framing. The chapter also picks up on my earlier focus on the problematic spatial framings of fatherhood with the suggestion that there is not only some ideological fatigue, but also there are some interesting surprises that push a more open politics that recovers the emotional act of fathering. Andy and Sue are adoptive parents whose emotional journey towards parenting paints an interesting picture of toiling beside (both with and against) medical science and its institutions. Finally, with this chapter I build on my earlier discussion on fathers' *becoming other* and I open Agamben's (1993) notion of *the coming community* as a political possibility. Andy helps me with this by illustrating his changes as a father, his changing partnership with Sue and the evolution of their communal household. I am also helped in this task by Michel de Certeau's (1998) later writing in favour of *orality* as a foundation of community:

---

1 If I embrace feminism as a powerful denaturalizing force then I must conclude that gender is a socially imposed division of the sexes, and this includes women's seeming natural relations to the birth process. As Audrey Kobayashi (1994, 77) points out, we need to fundamentally challenge anything that is meant to characterize "people according to criteria which may seem to present themselves as natural (because they have been naturalized) but are nonetheless based upon social choices."

> Orality also constitutes the essential space of community … there is no communication without orality, even when this society gives a large place to what is written … Social exchange demands a correlation of gestures and bodies, a presence of voices and accents, marks of breathing and passions … a visceral, fundamental link between sound meaning and body (de Certeau et al. 1998, 252).

Championing the re-introduction of more surprise to everyday spaces, de Certeau's (1984, 1992) first project focuses on the seeming naturalness of cities, communities and lived worlds, but it is equally applicable to seeming naturalness of parenting.[2] For example, the framings and discourses of medical science and its attendant institutions and bureaucracies (La Maz classes, birthing rooms, foster care, pre- and post-natal care, adoption agencies) remove large parts of the dynamism and flexibility of early parenting. By removing as much surprise as possible (e.g. with pre-birth knowledge of gender and genetics) they suggest life should be controllable and predictable. Alternatively, I draw on Doreen Massey's (2005) notion of *throwntogetherness* to argue that surprise and happenstance are part of the opening up of the political. I use Andy and Sue's adoption procedures, and their rebellion against them, as a case in point.

de Certeau's (1998, 251) second project (with Luc Giard and Pierre Mayol) focuses on culture as practised through orality, operations and the ordinary. Oral culture, he argues, "became the target that writing was supposed to educate and transform. Practitioners have been transformed into supposedly passive consumers. Ordinary life has been made into a vast territory offered to the media's colonization." de Certeau's point echoes Guy Bedord's famous admonition about the commodification of desire and fantasy, which I footnoted in the last chapter. With this chapter I want to move away from media representations for a moment and probe different ways of knowing the spaces of fathering from oral and material practice. In particular, I want to highlight the illusion of the natural father and favour, instead, the emotional work that makes fathering inevitable. This emotional work is part of becoming other and the coming community.

Oral Trajectories

Buddy is a young father. Full of self-dramatizing impulses, he is wasted-looking in a way that suggests a style rather than a set of depressing circumstances. He has a full face and the feral eyes of a boy absorbed in the task of surviving.

---

2 de Certeau's (1984) first project is to revitalize lived worlds by creating ways to recapture spatial stories and narratives so that happenstance is brought back as an important form of produced knowledge. He argues that narratives, spatial stories and trajectories are suppressed with the emergence of science (and particularly the biological sciences) as "the writing of the world" (Massey 2005, 25).

> Totally unready to be a father,
> I had not held down any steady job,
> arrogant,
> thought I knew everything,
> no experience with children,
> never even held a baby,
> no desire to be a father
> at the time.
>
> And … ah … something changed.
>
> Another life was going to depend on me,
> I started thinking about, …
> … wow,
> how much I wanted a daughter,
> 'cos daughters are always daddy's little girl
> and it just sounded so cool to me
> and I really started to like the idea
> and it motivated me.
>
> I work two jobs and take care of our lives.
>
> (Buddy, Cosmos Café, La Mesa)

Buddy's poetics suggest an inevitability that is not at all related to biological relativism, but to the work of fathering. His words – cropped from the interview transcript, edited and rendered in a different pentameter – are a poignant affirmation about the construction of fathering – as an emotional work rather than a right – that I want to highlight in this chapter.

I met Buddy on a desert camping trip when he was just turned 21. He was a father three years later. Diagnosed with Attention Deficit Disorder (ADD) and bi-polar depression, Buddy's psychological intrigues are elements he plays off his frenetic commitment to his two daughters. As he filled out over the years, I was still able to detect his early fearlessness and his schemer's flexible logic. Halfway through our coffee at Cosmos' Café, Buddy gets a phone-call letting him know that his step-father, Jim, who had been struggling with cancer, was dead. I knew Jim well, visiting him weekly at the hospital during the last month and a half. I knew that he was more real to Buddy as a father than his 'natural' father. Buddy races through conversations with ADD-powered fanaticism; whatever it takes to get to the edge. Jim's conversations were thoughtful and laconic. There is an orality to the relationship between Buddy, Jim and myself – a correlation of words, gestures and bodies – that marks a connection towards our coming community.

de Certeau and his colleagues (1998, 254) argue that "culture is judged by its operations, not by the possession of products … [that] communication is a *cuisine of gestures and words*, of ideas and information, with its recipes and its subtleties, its auxiliary instruments and its neighbourhood effects, its distortions and its failures." They go on to note that "[by] itself, culture is not information, but its treatment by a series of operations as a function of objectives and social relations." They divide these operations into three aspects. The first is *aesthetic*, and relates to everyday practices that open up a unique space within an imposed order (a spatial frame). They see this as a "poetic gesture that bends the use of common language to its own desire in a transforming re-use." The second is *polemical*, and relates to everyday practices and power relations that structure social fields and knowledge. What I try to do here and in the chapters to follow is take the information gleaned from previous chapters and, to put it in de Certeau's words, "bend its montage" to my own use through the words of fathers so as to take power over a certain knowledge "… and thereby overturn the imposing power of the ready-made and pre-organized." de Certeau's third aspect is *ethical*, and relates to everyday practices that restore "a space of play, an interval of freedom, a resistance to what is imposed." To be able to do this is to provide a Bergsonian dislocation, to establish a distance from what is imposed.

In what follows I use de Certeau's notion of operations to loosely interpret (and then let go of) a conversation with Andy.

"Fatherhood was Just Something I Expected"

Andy and I are sitting on the deck of my home, looking north to Cowles Mountain. Andy is a relatively young man, and his sentiment about the inevitability of fatherhood reminds me of the ways Buddy's poetics created a space for fathering. I sit with Andy during one of those interminable Californian summer mornings. The sharp aftermath of pre-dawn cold dissipated about an hour earlier and the promise of a balmy but not uncomfortable forenoon descends upon us. Andy looks wistfully at the mountain. Not so long ago he bought a house about 25 miles distant, on the other side of the mountain. It was a home space, bought in anticipation of parenthood.

Today, Andy has some work to do near where I live and as a consequence is happy to stop by my place for this interview. Andy likes to go out of his way to help other people, and it seems to me that dependability is etched in his face and body. His strong, aquiline features converge on a prominent Romanesque nose. It would be difficult to contest the measurements or nobleness of Andy's facial features, which rest fitfully on a tall athletic frame. Any semblance of sternness disappear with Andy's smile that embraces every feature on his face and then projects forward, outward to whomever he is conversing with in an almost conspiratorial fashion. It is not the kind of smile that embraces a room full of people but, rather, it contrives a fealty to the immediate listener. All these mannerisms come together in Andy to suggest trustworthiness and integrity.

There is also in Andy a sense of wonder and delight at the workings of the world. He exudes a joy that seems to acknowledge life's frustrations while projecting an acceptance. For all these reasons, I felt a tug of compassion for Andy and his wife as they struggled to get pregnant. In the end, after many years of fertility consultancy and other trials and errors, Andy and his wife decided to adopt. Andy – the most natural of parents – is not a 'natural' parent. I wanted to talk with him about this particular journey to parenting.

"When Scott came into your life earlier this year, you were catapulted into fatherhood," I note. "Tell me about the journey that got you to that place."

"Ah, well, it is a long story," Andy replies. "I got married in 1992, August 1992. So, you know, fatherhood was just something that I expected. It wasn't something I thought a whole lot about like: 'will or won't we?'"

Andy talks about his family of origin, which clearly demarcates aspects of how he understands his own fatherhood. He talks about his upbringing with his sister and then describes the closeness he feels in his communal household comprising himself, his parents, and Sue's parents.

"They're still married," he notes of Sue's parents. "That was important to me, and having kids was important to me. My wife comes from a really large family. She had …" Andy pauses, correcting himself, "well, maybe not really large, but four siblings: three sisters and a brother separated by like six, six and a half years. So her mom had them like back-to-back-to-back. They did everything together. I just kind of assumed that we'd get married and start having babies …" Andy's voice trails off, faltering slightly.

"Em, well … eh …" Andy looks at me pointedly (he and I have talked about this part before).

"And *you* know that really didn't happen for us."

An unpleasant memory.

"Actually, when we were dating Sue got pregnant and we decided to have an abortion. We were not real serious at the time and I really, really had some stuff to work out."

Not only was this the beginning of Andy's journey to fatherhood but it set a context – an imagined but nonetheless palpable emotional context – for his later difficulties with pregnancy.

"After experiencing what I experienced trying to have children later, I thought – some part of me felt like – this was my punishment. A little bit. You know: you had your shot and you lost it! You blew it! I don't know what would have happened had we had that child and what kind of father I'd have been able to be at that point in my life."

"What age were you then?" I ask.

"Eh, we started dating when I was 20 so I was probably 21. What I would have been like I don't know or what that would have turned out to be I am not sure. But I definitely think in retrospect that everything happens for a reason and it was the best thing for all parties. And knowing what our history was after that, [I have to question] whether she'd have been able to deliver the baby. I don't know. We've had several miscarriages. She was pregnant for 10 weeks at one point, which was really rough. She was getting excited and telling people, friends. We tried IVF (In Vitro Fertilization) twice in 2004. We made a visit to an IVF doctor five years previous to that. And I had a low sperm count so they said our chances of getting pregnant were pretty slim. Then we decided to revisit it and the technology had changed so much that my problem was pretty insignificant, but at that point my wife had a problem with her egg quality and some hormone levels. The two of us together – you know with my problem and her problem – the two of us made the perfect scenario for not having babies."

There is an interesting balance in this part of Andy's story suggesting what might be construed as ambivalence, as he talks about losing passion for his journey toward fatherhood. This is perhaps less about lackadaisicalness and more about some problematic complicity that confers on women an unasked-for authority over decisions about parenthood. I'll come back to this problematic a little later. Perhaps, more appropriately for Andy and Sue, the ambivalence is really about the relational complexity of parenthood. It seemed to Andy that technology had let him off the hook not just for his culpability around infertility, but also for a number of other things. Nonetheless, Andy is a compassionate man, and he remembers how hard this was for Sue.

"So it was a little odd me always feeling like I was, you know, I was the indicated problem in our relationship. And for the doctor telling us that my problem wasn't so significant and that it was my wife's problems that were going to keep us from having a baby," Andy pauses reflectively. "And that was a heavy blow for my wife. She took that really, really hard so it was definitely a challenge trying to stay together and she really, really wanted it and I was at a point where I was pretty happy with my life and had kind of come to a place where I could accept that we weren't going to have kids. We would have discussions all the time and it would be like okay lets get on with our lives together and travel and enjoy life and everything would be great. And it usually did not last that long. And we'd have another discussion and it would be back to trying to do something.

We talked about [using an] anonymous sperm donor; actually tried that twice … And she actually got pregnant and lost it really early which had been kind of our M.O. for our whole relationship …," another pause. "She'd get pregnant but couldn't hold it. That kind of led us to a point; you know, we'd talked about adoption for a long time and decided that that was something we wanted to do and a part of me just didn't think it was ever going to happen, you know? Knowing who I am and who she is …," Andy's emotions and reasoning around this issue are very much tied to larger contexts of his identity and his relationship to Sue.

"We're good at starting things and not finishing things … we have big eyes or big dreams: all of that sounds great we should do that, [but we] never have the motivation or the follow through to finish up and do it."

"By then you'd kind of accepted you wouldn't be a father?"

"Yeah."

"How did that acceptance come around for you?" With this question, I wanted to start to get into the particulars of Andy's fantasy about fathering.

"Eh, slowly, you know? It is just something I always thought I was going to have – a little boy. I always thought I'd have a boy that I'd be able to play ball with and do all those things: watch him play games and ride a bike with him and so all those things I love to do, or a girl for that matter. I don't know why, I just always thought I'd have a boy … Just getting to a place where I would accept that wasn't going to happen was difficult but I think em … I don't know, I had a really good life, you know? I was pretty happy with where I was and, honestly, a lot of our friends had younger kids: visiting them did not make it real appealing to have children, honestly!"

Andy's smile grows bigger as he remember a litany of reasons not to have children.

"So a lot of it was like maybe, maybe it is a real blessing. You know? We'd go visit them and they'd be pulling their hair out and oh my god its, you know:

> they don't do anything,
> they weren't getting along,
> they didn't have sex anymore,
> they didn't go anywhere,
> they didn't do anything,
> the place was a disaster,
> there was kids' stuff everywhere,
> they were ready to kill each other.

Part of it was so mysterious to me, I don't even know if I could be a father, what I would do, but just interacting with my nieces and nephews has been a real, real blessing and a lot of fun but also great when at the end of the day you know, you know, well they need a diaper change or they're starting to melt down, you know, starting to fall apart at the end of the day and it is time for them to go home and I get to go read a book or do whatever and they've got to deal with it. So I guess this selfish part of myself was really happy with my life and contented and just kind of rationalized it that way.

My wife wasn't the same, she was a little more tortured than I. I think it is that motherly instinct that women tend to have, you know they feel maybe incomplete if they don't experience that.

And then coming around to adoption was difficult for me. I had really mixed emotions about a lot of the [stuff]. The anonymous sperm donor thing was hard for me because it felt like I'd be looking at this child who wasn't mine. And what that would feel like: would I feel less than or feel like I wasn't really a part of… And then the adoption: that wasn't really our child and how would that make me feel or, you know, somebody saying he looks just like you or throws the ball just like you or he does some mannerisms like you or whatever and I always felt like I'd be saying to myself, if you knew the truth you wouldn't, you know, that I would feel like that on the inside …"

Andy and Sue went the adoption route and on this particular day, as we sit on my deck, they have just finalized the papers to adopt a three-month-old boy. Their struggle to adopt is long and convoluted to the extent that I want to gloss over some of the details. Rather, I want to focus on Andy's feelings as a father. I want to pick up the conversation where he brings up the notions of natural fatherhood and the mannerisms that children, supposedly, are genetically inclined to gather as part of their biological make up.

"And what's amazing is that it has been absolutely 180 degrees away from that, you know? He does things that are like me or his nose looks like mine or he looks just like his Daddy. There is no weirdness about it at all, which really blew me away! Just instantly, when I met him – which was very surreal: we met at a Denny's in Ramona and just the circumstances of sitting there and waiting for your child to come in … My son is about to walk in and be in my life now and [I'm] sitting in a booth and waiting for that to happen. [It] was just really, really odd. Just a magical moment, but just very surreal. And, I mean, as soon as I saw him it was like 'that's my boy'. That's him and it was … I cannot even describe it … It was just totally … It felt fated, like it was absolutely supposed to be and all the stuff that led up to where we were … It couldn't have happened any other way. It was supposed to happen that way."

As he talks, there is a lyrical aesthetic to Andy's words. And as he relates his experience to me orally, there is a sparkle in Andy's eye and the smiles on both of our faces grow huge. Poetic experiences like this – the one's that are indescribable and are total – are the essence of the work that describes emotionally what it is to be a father. There is a rightness, a clarity, a completeness to Andy's words/poetry as he describes his son – someone else's son/his adopted son – arriving at a booth in Denny's restaurant in Ramona.

> I met him
> Surreal
> We met at a Denny's in Ramona
> Sitting there and waiting for your child
> to come in.
>
> My son is about to walk in and be in my life
> Sitting in a booth and waiting for that to happen.
> Really, really odd.
> A magical moment,
> surreal.
>
> I saw him
> like 'that's my boy'.
> I cannot describe.
> It was just totally
> absolutely
> supposed to be.

This is the inevitability of fathering. For Andy, this moment and this space are the culmination of a myriad of conflicting emotions that circumscribe a rugged terrain – an initial abortion, numerous failed pregnancies – and the journey that Andy and Sue and their communal family embarked upon years earlier.

Throwntogetherness

When Andy describes his first meeting with his son I am reminded of Doreen Massey's (2005, 149–52) use of the term *throwntogetherness*. The term sounds vaguely Heideggerian in the sense that it suggests becoming and belonging, and there is certainly a contrived naturalness to the way Andy describes the instant connection that he felt for Scott at *Denny's*.

The poetics of heart work, I aver, are where a power over the politics of otherness finds a form that is material and geographic. And yet, in describing a material and geographic connection with *throwntogetherness*, Massey is suggesting "the politics of the event of place … [P]laces pose in particular form the question of our living together" and the problem with so-called spatial politics is that they are concerned with how the "irreducibility" and "instability" of space can be ordered and coded, "how the terms of connectivity might be negotiated … Just as so many of our accustomed ways of imagining space have been attempts to tame it." What happened for Sue and Andy at the Denny's in Ramona is not at all about the contrived space of the restaurant or the spatial framing of the County Adoption Service but about the space of fathering and mothering, their trajectory, the trajectory of Scott and his birth mother, and his foster parents. It is about halting, uneven, and insubstantial trajectories that create avenues for getting beyond traditional spatial frames demarcated by institutions, bureaucracies and hierarchies. It is a subtle and flexible mapping that recognizes the power of affect.

Taking stock of emotions and the interdependencies that come with any relationship, Andy continues thinking about how he came to fathering and that throwntogether moment in Denny's. What follows is Andy's description of the months of trials that led to this moment. It is a series of thoughts that foments a mapping that leave no doubt in his mind that he is Scott's father.

"The whole IVF thing was really, really hard on our relationship … my emotions. You know, I would hold back and hold back and try and feel and Sue would get upset with the way I was dealing with it because she was excited and glowing and talking about where he was going to go to college, you know, and it is like I can't go there right now, I can't get that excited and she, you know, she would get mad at me because I was being negative and I was like I am just protecting myself and then, you know, I guess I was just more even-planed than Sue. She was really up and down and I couldn't go up with her and I didn't because I didn't want to go down with her, but I still, you know, I still did a little but not as drastically as she did and that was really, really hard on us and so it was hard for me when she said 'Do you want to do the adoption thing?', and I'm like, 'Oh it is just one more thing that won't work out somehow or something will happen or it'll just get dragged on, we won't follow through with it'. So, I don't know, part of me was reluctant and the other part was, well, just start doing the foot-work and let go of the outcome."

The emotional work of fathering hinges in part on relational dependencies – on past models of family and fatherhood that we internalize – and, in part, on the happenstance and the unexpected. Emotions are more often than not unexpected and yet they bear heavily on how I come to fathering, how I invest in (or not) differing relationships with my children, and how flexibly I move through the unexpected events of those relationships. The questions that are raised for me from Andy's complete, instant and natural connection with Scott foment around what constitutes biological connections and relations and how those connections are trumped by emotional bonds.

Institutionally Framed

Fostering and adoption is an institutional quagmire that exemplifies an attempt to frame the context of parenting. Andy helped me to understand how his connection with Scott is much less about this formal, traditional framing – the "set of bureaucratic tracings" – and much more about a different kind of mapping, a tender mapping, that is a journey whose ending is much less known than dreamt.

"How did you choose an adoption agency?" I ask.

"We actually just chose the County because we didn't have the money to do anything else. It was financial, honestly. With the County everything was free, all the classes were free."

Classes. One element of the spatial framing.

"Is this a state-mandated program that you have to go through in terms of the classes?" I ask.

"Yep, yep. Yeah, the County of San Diego, the State of California. So it is children that, that …," Andy shifts laterally to thinking about the machinations of state adoption processes. "And that was another thing that we had to come to accept: they pretty much smash the idea that you are going to get the *Norman Rockwell* type of child or the idyllic, perfect, healthy, smiling, bouncing, little, blonde-haired, blue-eyed baby boy."

"How'd they do that?" This is interesting. Not only are Andy's preconceptions of natural fathering challenged, but so also are his preconceptions of the characteristics of his child.

"Well, the fact is that most of the children that come through the system through the County are being taken away from their parents for drug exposure, child endangerment, molestation … You know, exposure to drugs or they test positive to drugs when they're born or their mother does – abuse, neglect, all those types of things. So none of them get there on a good day. The amount of their exposure or damage that is done to them is kind of dependent on how old they are: if it is an infant, how much drugs they were exposed to when they were *in utero*. But none of them get there on a winning streak. So they are trying to tell you basically that the kids you are adopting are a little bit damaged goods and it takes some TLC and some special nurturing to build these kids back up. You can have a beautiful child but you are not getting, you know, your so called perfect little child. Most of them are gonna be African American or Hispanic: It is just the statistics with the children they have in the system. There's a glut of kids from three to teenage. It is really sad, they are less desirable and there's more of them and the older they get the harder it is for them to get adopted because they kind of get caught up in the system and bounced from home to home and then they start exhibiting behaviours and, it is a vicious cycle, they wind up having children that end up in the system or they end up in institutions. It is just a staggering amount of kids that are in the foster care system that wind up in jail; something like 80 percent wind up in jail … So I mean it is really bad odds. So that is part of it too, people do not want to take that chance so they adopt out of the country."

"And none of this dissuaded you and Sue? You just kept on going?"

"Yeah, we just kind of felt like, I don't know, part of me felt like maybe that was the child that we needed to have because of our history and my history and that maybe was the path that we were supposed to take."

"Does the County system assign you a caseworker?" I ask.

"Yeah."

"And you give your preferences for the kind of child you want?"

"Right."

"And she or he is looking for that?"

"Right, right, that is a bizarre thing. You have a form and you are checking boxes of age range, race or ethnicity, preference on gender, and then what you are willing to accept in terms of sexual abuse or physical handicaps, mental handicaps, drug exposure. And then [it gets] specific: alcohol, severe alcohol, em you know fetal-alcohol effect, fetal-alcohol syndrome."

"Tell me about the education you are getting as part of the classes you were taking?"

"PRIDE was one of the classes we had to take."

"Is that an acronym?"

"Yeah, it's Parents Recognizing … er, I don't remember …"

"Are you getting classes on regular parenting as well?" I cringe inside at my use of this particular adjective.

"Yeah, most of it's geared for adoptive parents. The PRIDE classes are for adoptive and foster parents and technically basically you have to get licensed as both because you are a foster parent during the first six months, there is a six-month kind of probationary period where the child is in the system. Typically what happens is the mother loses her rights, the court is fighting to get her to lose her rights to the child and place the child in adoption so during that process there are foster parents and they are going to court and they file these petitions basically saying the mother should lose the rights to the child. The family or the father or the aunt or the grandmother can fight to keep this child as well. So it just kind of depends on that whole process and so it could be really drawn out and long. They've got some time, and then you know they need to jump through a lot of hoops so – going in recovery, supervised visits, and talk to a social worker and get counselling. It usually doesn't happen. Putting all that foot-work in front of them is usually too daunting even if they want it so it usually doesn't happen, but it drags out the process. We got really fortunate because Scott's mother relinquished her rights immediately and decided she was going to give him up openly."

Scott was born to a young woman who had shown evidence of opiates and meta-amphetamines in her system. She was very young and hid the pregnancy from her family and the child's father.

"After he was born and she was confronted, she was like 'I don't want anything to do with him' and then she had a social worker that came in from the County that talked with her. She was basically the child's social worker/caseworker and said, 'This (the adoption) is something you should be a part of: there are things you can do for him and you can choose a family. You can be a part of the process and it will really help you in the healing part of this if you are involved instead of just saying 'I don't want him, I'll sign right now and I'm out of here.' So they really encouraged her to be part of, to give her family history so it would help them with any medical things."

"In the meantime are you getting options on a bunch of babies?" I ask.

"We hadn't gotten any yet. We got licensed in December and we had been just sitting around. It was February 10th, I think, when we got the phone call. So yeah it was an amazing moment getting that phone call."

"Yeah, did you meet the birth mother before?"

"No …"

"So it all came together in that moment." A throwntogetherness, the merging of stories, of trajectories. Indeed, as Andy's next set of comments suggest, the process was much more happenstance than is prescribed by the careful framing that the County would prefer.

"Yeah, we jumped the gun a little [when] we met the child at Denny's with the foster parents. The mother never had the child at all, he went from the hospital to foster care in Ramona actually – which is where I live – so you know he came from Los Angeles, the mother lived in Los Angeles, she gave birth in Encinitas because her father lives in Carlsbad." Geographic serendipity.

"So, because she came here he wound up in the system in San Diego County as opposed to LA County whereas if she would have had him the day before she came down here (on Christmas Eve instead of Christmas Day) he would've been in LA County foster care system and I wouldn't, she would have never seen our profile." Temporal serendipity.

"His birthday is Christmas Day?" There are layers and layers of providence in this story that, like all births, are miraculous. Andy's next comments nail the notions of happenstance, of fate, of surprise, of powerlessness, of throwntogetherness.

"Yeah, yeah. Christmas night. Yep, so a late Christmas present. So, yeah, so just the way everything kind of worked out it really … You know when you are doing something it, it feels like you are forcing it to try to happen; it always felt that way with our trying to have a child, that we were really trying to impose our will on, on making it happen. And it really felt different with the whole adoption thing, you know, we were just doing the foot-work and, you know, you get licensed and they say okay we believe that you can parent this child and you just literally, you just fill out the forms and you just let go, and you are done. I mean there is nothing for you to do other than for them to say 'Hey we have this child for you, do you want to consider it or not?' So you really, really kind of just do the footwork and let go, I mean absolutely. So, em, so the way everything just came together … it really felt like it was supposed to be. There were so many things that were just amazing. Just the odds of them happening were so slim it just seemed like it was absolutely supposed to be."

"Who was all there at Denny's?" For me, this is the apex of the throwntogetherness in the sense that it precipitates a huge emotional push for Andy.

"Ah, Sue, myself, em Scott – my son – and the foster parents."

"The caseworker?" This is where the framing, the institutional power, falls away. This is a transgression, a resistance, a flagrant dismissal of an operational and legal framework. And it is a push that comes from Sue.

Pushing Around the Institution

"No, actually – it was a Friday night. We actually weren't supposed to meet. It was the day after we had our *telling*. They have this thing called the 'tell' – which we had at the Health and Human Services Office with our social worker and Scott's social worker and the woman who'd been counselling the mother – and they pretty much tell you his whole history so everything that they have – medical background on the parents, background on the parents' parents, her whole history, how she grew up, everything they can glean from her – they tell you openly and then basically you make a decision whether you want to adopt this child, knowing what you know, do you want to proceed? Because they don't want you to go 'well, we'll see, we'll meet him, we'll get involved, and then you change your mind.'"

This surprised me.

"You gotta make a decision before you meet the child?" I asked.

"You have to make a decision whether you want to proceed, yeah!"

"What does that mean, 'proceed'?"

"Well they want you to make a decision whether you want to do this or not so you can still – you are not locked into anything but you are basically making the decision whether you want to do it or not, em, because it is harder, especially with older children … But yeah, we were in all the way, we saw that what was, was just a perfect scenario. We were really, really fortunate with just the way it was going down … Usually the kids come into the system, the parents have been in jail and institutions and drug rehabs and arrests and you know, their parents, it is just a … that is what you are getting … And that wasn't the case with him at all so it was just all, the whole scenario was just like one in a million for us so we said yes right away.

"They told us where he was in foster care and we were just blown away, you know? He was in Ramona, which is six miles from my house. And we asked if we could meet him and the mother's social worker and Scott's social worker said, 'I don't see why not' and she was looking at our social worker and he is like 'No! No! I don't want you to.' And we are like 'Why not?' 'I don't want you to get any more attached to this kid than you, you are,' he says. 'It is gonna be bad if she ends up changing her mind or something happens.' We were supposed to meet her, the following Thursday, so a week later. And then we were going to take him home hopefully like the day after that or so. So he tells us we can't see him and I'm bummed, but okay.

My wife wasn't having it, she was just so upset: he's in Ramona and she's sitting at home, she works from home so she's sitting at home, can't think of anything else but the fact that he's ten minutes away from us right now. We got his whole file at the telling so we had the foster parent's number and everything so they had encouraged that we actually called them and talked with them just to get his schedule – what he does, when he likes to sleep, you know, does he like the room light, does he like it dark, does he like to be sung to, does he like noise – you know, all those things that would be helpful to us. So my wife started speaking to the foster mom and she was very sweet and oh, you have to meet him, you know? They both basically decided that our social worker's decision for us not to meet him was a poor one and I was like, 'I don't wanna, I don't think it is a great idea either.' My feeling was, dude, we are already sucked in emotionally on this thing and if you tell us it is not going to happen I am going to be devastated. Seeing him is probably going to make it worse if it fell apart but I am already going to be devastated if it doesn't happen so you know you're not sparing my feelings from keeping me from seeing him, I am already emotionally invested."

"So I am curious, you mentioned earlier that you were kind of reticent to push the fostering, but clearly something changed for you in this process, can you articulate what that change was?" I am thinking about the poem from Buddy earlier in the chapter, and his emotional change when his girlfriend told him he was about to have a daughter.

"I really felt like that was going to be our path, like that was what was supposed to happen and part of it was just I really felt like we were doing something good. You hear these heart-breaking stories of these children and not only that, but the parents you know, the parents and what their experience is and how hard it is for them to lose a child and what that must be like and you see these videos …"

I interrupt. "So the whole process, the education, the caseworker, invested you in doing this?" Could it be that institutions such as the County adoption and foster care system are both illness and remedy?

"Yeah, the PRIDE classes for sure, I think it was seven weeks long and it was a whole day, Saturday from like nine to three or something. It was really well done and the people that taught the class were social workers actually. Our social worker wound up being one of the PRIDE teachers and, amazingly enough, when we bought our home, the people who we bought our home from had adopted through the County. They had a 6 month old baby girl and we told the social worker – the guy who was teaching our PRIDE class, the guy who turned out to be our social worker – 'it is funny the people we bought our home from they adopted through the County.' He went, 'Oh yeah, what's their names?' And we were trying to think of it: Gilbert, Gilbert that was their last name and he goes 'Oh, Paul and Penny' or something like but eh it was funny." Serendipity within serendipity. The everyday is a continual happenstance throwntogetherness.

"So let's get back to Denny's and the illicit meeting," I suggest.

"Yeah," Andy laughs. "I was saying 'No, I don't think we should', so she'd go 'Okay, yeah, you're right' and then she'd call me again and go, 'What do you think? Do you think we should?' You know, she wasn't asking me! So I told her six different times you know, 'I really feel strongly that we shouldn't do it.' And the problem was that the social worker was off on Fridays so she had thought about it and talked to the foster mom Thursday night and then again Friday so we couldn't really get a hold of him to say … you know? And I said if we could talk to him and plead our case and say 'Here is how we feel about it, what do you think?" I was all for that. Call him and tell him this is how we feel and we feel strongly that we should see him and that we are already emotionally involved. So she left him a message but he was off Friday, Saturday, Sunday so unless it was an emergency he wasn't going to get it until Monday. So, she just, she couldn't last, she couldn't make it. So the baby's social worker had spoken to the foster mom and basically given her a heads up: 'Sue might be calling you and its okay to talk to her about Scott.' And Sue called the foster mom and said 'Ramona's a small town so if we happen to bump into each other at eight o'clock at Albertsons or at the grocery store,' you know? So they just took that and ran with it. They lived just behind the Denny's so they are like, 'Let's meet, he usually takes a nap and he is up from eight o'clock 'til ten or ten-thirty.' And so my wife set it up so she said, 'We're meeting at Denny's at 8 o'clock,' and I am like, 'What are you doing?' And she said, 'Well that is it, I'm meeting him, and if you don't want to go and meet your son then fine and I'm going.'"

Andy starts laughing.

"Oh god, I'm dammed now either way 'cause she is gonna hate me for the rest of my life because I didn't want to go meet him. And it wasn't that I didn't want to go I just really felt strongly 'cause he – the social worker – felt strongly about it.

So, we went to Denny's. It just worked out wonderfully."

For the next half hour, Andy tells me about the bureaucracy involved with fostering and then adopting Scott. He goes on at length about their benign reprimand from their social worker and the forms and meetings and anxiety that went on before Scott's birth-family finally signed the release papers.

Last Sunday I stopped off at Andy's house for lunch. He and Sue were throwing a party to celebrate their promotion from foster parents to adoptive parents. Seven months have passed since that eventful evening at Denny's and the prescribed six months of foster parenting are now over. Scott is scooting about on his butt chasing a small ball. He can stand and walk but is in that space where he feels much more confident, and faster, on his butt. He loves playing catch so I scoot down on my butt and we pass the ball between us for a while. Then Andy's mother comes over and I am out of the picture. Scott deftly rolls the ball to her while pausing to give me a big smile. With that smile, I don't feel at all put out by Scott's decision to move the ball in Grandma's direction rather than mine. It is exactly his father's smile, the one that brings a near acquaintance into a confidence.

The Nature of Fathering and Other Awkward Emotions

Andy's story elaborates de Certeau's (1998) notions of *orality* and *operativity* through everyday practices of aesthetics, moving beyond polemics and institutional frames to create an emotive ethical space that is simultaneously a dislocation and a re-construction. It highlights a strange immiscibility between fathering and child-care that I am not quite sure how to reconcile. Can it be that part of the problem relates to men's inability to bear children and provide them with 'natural' sustenance as infants? To suggest such is to return to a form of naturalism that is groundless in the light of contemporary feminist and post-structural critiques. It is also groundless in the light of Andy's experience.

Thomas Laqueur (1990, 1992) points out that the "biological facts" of motherhood and fatherhood are not "given" but come into being as science progresses and sexuality assumes cultural significance through political struggle. This is precisely what Massey (2005) is getting at when she talks about opening up space as more than a stage for that struggle but rather as an integral part – a production and complicity, if you will – of that struggle. Medical science does not biologically construct the work of fathering. The problem, Massey avers, is that with spatial frames and traditional mappings (such as those that are written by science), the political will to identity formation is foreclosed upon. Nor do the issues that contextualize family and fathering cohere around the spatial frames of patriarchy described previously. The contexts of fathering in families is about action and doing and is, in its fundament, an ethical issue.

Contemporary critiques of empiricism suggest that no fact – such as the ability to bear and nurse children – entails or excludes a moral right or commitment. As fathers, then, Andy, Buddy and I are comforted and liberated by the idea that gender is a socially imposed division of the sexes, and that there is no 'natural opposition' between men and women. But current fatherhood ideology and social science research continues to define a father's relationship with his children primarily as co-parenting that is interdependent with, and sometimes in opposition to, mothering.

I want to continue Andy's story with another ethno-poem that is, I think, poignantly directed to fathers. At the very least, it resonates with my experience.

It is another pivotal moment for Andy, perhaps as emotionally charged as the one in Denny's. It is his first day alone with Scott. Andy describes for me some of this oppositional tension and how it plays out in his relations with Scott and Sue.

> One of the coolest days
> to spend the whole day
> with him
> by myself.
>
> Am I going to be able to do this?
> What if he cries and he won't stop?
> What am I gonna do?
>
> I was definitely involved in changing diapers and doing that thing,
> but it was like …
>
> (Sue saying),
> 'Here let me have him,
> he's not responding quickly enough to you,
> let me take him'
>
> You know?
> I was still kind of a second class citizen.
> I think she was a little sceptical.
>
> How was it gonna go when I was alone with him?
>
> You know?

I know! Andy's reservations and their attendant emotions remind me of a pivotal moment for me with my son. It is a moment that began my thinking about the awkward spaces of fathering (cf. Aitken 2001).

Shortly after Ross was born we were visiting my partner's family in Canada. At a large family gathering Ross began to fuss and I picked him up and laid him against my chest to quiet him down. I had been comforting Ross since he was born and only a small part of me recognized that this was a more public event than I had heretofore experienced. Needless to say that same small part of me wanted to show off my newly honed fathering skills. This time Ross would not be quieted. As time progressed painfully slowly and Ross' discomfort found new heights of vocalization, heads began to turn with looks that mixed sympathy with what I thought were smirks. Ross went from my chest to my shoulder, then the other shoulder and then onto his stomach in my gently swaying arms. He continued to protest and the agitation in my gyrations and swings increased. Finally, my sister-in-law smiled sweetly and extended her arms towards the squirming bundle of baby. In frustration and with a sense that something insidious and beyond my grasp was occurring, I gave Ross up to her. Of course, he stopped crying immediately. My sister-in-law turned to me with an even more sickeningly sweet smile and cooed – half to Ross and half to me – as she laid his head between her breasts.

"Your Daddy just doesn't have the necessary accoutrements."

To be bested in this way by my sister-in-law, a staunch supporter of traditional nuclear family values and a vehement anti-feminist, was too much. How could Ross do this to me? Why was my frustration turning to outrage? Why was I over-reacting? I folded my bony arms around my hard, breastless chest and smothered my anger and shame so as not to show the pain that might explode and ruin the rest of the evening.

Faltering Performances

To think of a fathering performance as it is specified by Judith Butler (1993, 2) as a "reiterative and citational practice" becomes problematic because it inflexibly positions the ways that the daily conduct of fathering and the culture of fatherhood relate to gendering, aging, domesticity and so forth. If I accept that the performances of caregivers are amongst the most influential in the evolving political identities of young children, then it is important to study those performances. How, then, can I characterize the performances – and particularly the gendered performances – of fathers? In my past writing on families I suggest that although the day-to-day practices of mothering and fathering may be changing, there exist overbearing covert cultural norms that continue to be bound by an insidious patriarchal logic (Aitken 1998). Here I try to move beyond this by raising concerns that research on 'nurturing men' is misplaced precisely because it does not account for the space and work of fathering in relation not only to the norms that continue to define 'fatherhood' but also, neither does it account for the evolving father *becoming other*, connected to a communal household.

What I want, and what this chapter moves toward, is an idea of fathering as an emotional, poetic work. This form of fathering always trumps the fact of natural motherhood (although not, of course, the idea of mothering as an emotional, poetic work). In the next chapter I focus more fully on fathering performances, reaching and stretching with Butler's theories to an affective fathering and *becoming other*.

What I am left with in this chapter is a return to de Certeau and his colleagues' practices of everyday life (1998). The body of this chapter runs with their notions of *orality* and *operativity* as aesthetics, polemics and ethics. Ordinary culture, they argue (1998, 256), "hides a fundamental diversity of situations, interests, and contexts under the apparent repetition of objects that it uses." Difference, diversity and pluralization, they point out, are born from ordinary usage. The problem is that we do not know well the types of operations at stake in ordinary practices because our instruments and analyses are poorly suited for uncovering the complicated non-representational and emotive renderings of the everyday:

> In its humble and tenacious way, ordinary culture thus puts our arsenal of scientific procedures and our epistemological categories on trial, for it does not cease rejoining knowledge to the singular, putting both into a concrete particularizing situation, and selecting its own thinking tools and techniques of use in relation to these criteria.

To get beyond this blockage, de Certeau and his colleagues return to his first project to argue that we (you, me, those who are immersed in the throwntogetherness of the everyday) continue to work with "the innumerable ruses of the 'obscure heroes' of the ephemeral, those walking in the city, inhabitants of neighbourhoods, readers and dreamers … [that] fill us with wonder."

For me, Andy and Sue are those heroes. We saw that their orally constructed communal household held together through a series of trials, they pushed against the operations of the County's adoptive agency, and, through a series of everyday serendipitous events, they bore Scott, the most natural of sons.

Chapter 6
Stretching the Imagination

When Thomas Laqueur (1992) argues that the 'biological facts' of motherhood and fatherhood are not 'given' but come into being as science progresses and identities assume cultural significance through political struggle, he is suggesting the same kind of politics that Michel de Certeau and his colleagues (1998, 256) argue when they champion the "obscure heroes of the ephemeral." There is always a struggle between everyday cultural and a science that frames the world in a particular way.

Science often creates a biological imperative on what constitutes natural fatherhood and motherhood and this then is reified in the courts. Through this conduit, the notion of emotional bonds determining rights of parenthood flies in the face of legislative motions. In the UK, a recent legislative move attempts to sideline fathers from raising babies born through In-Vitro Fertilization. The 2008 Human Fertilization and Embryology Bill limits the legal duty on doctors to consider a child's need for a father before giving IVF treatment. The core of the medical science in the debate is the ability to create hybrid embryos that are a mixture of human and animal (0.1 percent) tissue. These embryos allow study of genetic defects and the cells formed in the process could help cure diseases. The core of the legislative discourse in the debate is that existing UK law requires IVF clinics to consider the "welfare" of any child created, and this currently means considering the need for a father. The new Bill removes this stipulation. The argument is complicated by new wording that designates "supportive parenting" rather than fathering. Under the current law some lesbian couples and single women in the UK are not allowed treatment. Opponents suggest that in practice, questions are rarely asked of lesbian parents and the larger issue is the denigration of fatherhood. A survey by the Centre for Social Justice, a think-tank chaired by former conservative leader Iain Duncan Smith, found that "80 percent believed a child has the right to two parents, and that 60 percent say those should be a mother and a father" (Lea 2008, 10). Most agree that the debate is semantic and symbolic, but nonetheless it raises a moral dilemma on the rights of a child to an adult man as a father.

In a different political arena, a similar issue arose in a 1999 US Supreme Court decision favouring mothers in child custody and citizenship disputes (Nguyen vs. INS, 99-2071). Like the UK controversy, the decision to accord mothers more rights than fathers when a child's US citizenship is disputed raised larger issues of who is a more 'natural' parent: "The difference between men and women in relation to the birth process is a real one," said Justice Kennedy concerning this legislative move, "and it provides a reasonable basis for the law" (*Los Angeles Times*, June 12, 2001, A9). The position of the US Supreme Court suggests the natural right of motherhood, and presupposes that motherhood is ontologically different from fatherhood. Proponents of greater rights for fathers such as Laqueur argue alternatively that no fact – such as the ability to bear and nurse children – entails or excludes a moral right or commitment. His point is that if this ontological difference between mothers and fathers is a reasonable basis for law, then the idea of fatherhood can never triumph over the fact of motherhood.

It seems to me that the position of the US Supreme Court ruling and the UK Human Fertilization and Embryology Bill fly in the face of contemporary academic discomfort that immutable qualities exist on the grounds of sexual difference, and that these qualities are natural rather than socially constructed. Laqueur offers an alternative in an extension of his earlier opus on the social and biological construction of sex: he calls it a "labour theory of parenthood in which emotional work counts" (1992, 155). As suggested in the last chapter, the notion of fathering as an emotional work challenges the naturalness of fatherhood (and motherhood) while celebrating men's (and women's) emotional connections with their children, their communal struggle and their moves towards becoming other. With this chapter I move the notion of emotional bonding and work further by suggesting that the work of fathering is not necessarily gender-bound. Cindy helps me with this task.

Naming the Father

Steve Kroft, host of CBS' *60-Minutes*, is interviewing Cindy and Robin on a show entitled 'Family Ties'.

"This is a little unusual," he says, "I am still trying to get my mind around it." He smiles wistfully, pauses, and brings his hand up to stroke his chin.

"This is not a traditional family in any stretch of the imagination," he says.

The camera switches to Cindy and Robin, sitting opposite Kroft. Robin is smiling and nodding her head. Cindy asks, "What is a traditional family today? I mean, I did not have a father growing up."

Kroft's show is focusing on families that use anonymous sperm donors; as part of its representation of "family ties," *60-Minutes* seeks to understand how a number becomes a father. Robin was artificially inseminated by donor 48QAH from a fertility centre of California. Wade was born in March, 2003. The letters QAH are memorable because the nurses at the clinic label them "Quite a Hunk" to characterize the donor, Matthew Niedner. Niedner regularly donated sperm to help finance his way through medical school. Robin and Cindy found Neidner through a chance meeting with a woman who had a baby from the same fertility centre and who also remembered QAH. They realized they had a sperm donor in common and that their children were biologically defined as siblings. They then went on the web-based Donor Sibling Registry, the mission of which is to bring siblings of sperm donors together. Using registries of this kind, donors like Niedner opt to give up their anonymity. The decision, for Niedner, was to help parents and children get more information about their biological roots than was afforded by a brief profile card from a fertility centre that legally respected donors' anonymity. In this sense, Niedner is more than a number. And he is not Wade's father.[1] Cindy is the father.

The *60-Minutes* special focused on the ways that sperm-donor siblings come together to create complex family connections. Wade, for example, is now connected with his half-sister, who lives with her single-mom in San Diego. The story I want to follow in this chapter is not about the *60-Minutes* focus on Niedner's fathering, but rather on Cindy and her emotional struggles with parenting.

The Girl-Daddy

Cindy is not easily domesticated. More than the strategically untamed hair, her penetrating blue eyes help Cindy retain a wild look from her days as a punk-rock bass player. I see in Cindy a cross between the hard angst of Joan Jett and the aloof cool of Deborah Harry. Cindy and Robin met in those heady days of gigs and clubbing, enjoying an unsettled lifestyle of which children were no part. They were romantically involved for over 20 years before deciding to have a child. Given her experience of growing up fatherless, it took a long time for Cindy to feel comfortable with parenting as an appropriate option.

"We were together for 10 years before we started talking about it," she tells me. "It's always been Robin's dream. And she said, 'I always knew it was my dream to have children. I always knew I was going to have children.' But then you're gay and for me, well, I'll tell you why it wasn't my dream: I love kids but I knew I was too selfish. To me it is a huge responsibility and not one that I would take lightly. So I said no for a long, long time.

"We've been together going on 21 years. So, you know the one time I said yes that was it, so the process started. And I was really excited and I was also scared but it was really her dream and I was really afraid. You know, and she is pregnant and I am 'oh my god I don't think I am going to be able to live up to this.' Am I going to be able to show up for this kid? [I was] afraid that I wasn't going to be good enough. And then the baby came and it was great."

Cindy and I spent an afternoon at one of San Diego's southern-most beaches. She was going to teach me how to wave-cast a huge, twenty-foot fishing pole into the surf. We sat in her motor-home looking out on an azure Pacific Ocean. It was a perfect day to surf-fish, but then we started chatting, and we never got to the surf-

1 In an interview given to *People* magazine on June 5, 2006, Matthew Niedner reported that "I do not consider myself to be these kids' father, but that doesn't mean I cannot develop a relationship with them and their families." He likens the sperm donation to blood or bone-marrow donation although he acknowledged in *USA Today* (June 15, 2006, 7D) that "sperm donation is a more complicated form of altruism." His marriage produced a child in the traditional way and he plans to "remain distant but available" to Wade and the other children in San Diego for whom he is a biological father: "I am open to a mature, friend-type relationship with these kids, if they are even interested."

casting. I've known Cindy since before she became a parent and the sperm-donor story exploded on the news. The spectacle of the news-pieces focused primarily on the sperm donor and the context of complexly mixed biological family ties. Niedner's space of fathering is particular, distant and limited. Cindy's parenting is connected to emotional work that stretches back to her relationship with her father and that is what I wanted to hear about.

I was also intrigued by the gender performances that she and Robin assumed around Wade, and I was here because of a story that Cindy had related to me a year or so prior. The story accords with something I've been thinking about since reading Judith Butler's (1992) ideas on how subjectivities are constructed and politicized by naming and repetition, and how subjects 'become other' by facing the power exerted upon them while, at the same time, assuming power as an instrument of their own becoming (Butler 1997, 11). To name a father or a mother, I suggest, following Butler (1992, 15), is to "call into question and, perhaps more importantly, to open up [the] term, like the subject, to a reusage or redeployment that previously had not been authorized." By so doing, strategies are embraced for thinking and feeling in new ways. As with previous chapters, I am raising the strengths and staying power of patriarchal fathering while at the same time contemplating its destabilization. Butler's notion of subjection (the creation of subjects) enables me to see how fathers are simultaneously 'made' and 'making themselves' in and through discourses and practices of institutions; their procedures and their recognitions. Cindy's story exemplifies the notions of a father 'made' and 'making herself'. How fathers become something different from the patriarchal notions embedded in dominant discourses is what interests me here. Becoming other is an active process that is never quite successful because the other is never attained. It is nonetheless an embracing of a process that is part illness and part remedy in a journey forward that is constituted emotionally.

Gibson-Graham (2006) bring together Butler's notion of subjection with William Connolly's (1999, 196) ideas on the "profound ambiguity within Being" to unleash the emotional tension between being and becoming. Similar to Deleuze's focus on affect and becoming other, Connolly (1996, 148) proposes a "visceral register" that is non- or pre-representational. It creates "intensities below the reach of feeling" that give rise to affective responses, gut reactions and embodied actions that are hugely influential in ways we become other (Gibson-Graham 2006, 24):

> So while Butler suggests continual pauses in the performativity of discourses and subjection that can offer openings for new becomings, Connolly offers a speculative glimpse of the way the new identities (and subjections) might arise to occupy these discursive aporia.

What I want to get to with this chapter are the ways fathering identities are changed, the ways those changes are remembered and emotionally prescribed, and the processes of naming that are the part of becoming other. Cindy helps me work through these issues with a story about a very specific gender-transformation.

"So tell me the story again," I urge Cindy, " just for the record, when Wade called you Dad, you remember?"

A big smile breaks across her face.

"Yeah, yeah, he, he said, …" Cindy pauses to make sure she get's the memory articulated to her satisfaction. She starts with the place. "He was in the bath tub and Robin was in there and he said, 'Mummy, tell me again why I don't have a daddy?' You know, 'Who's my daddy?'"

"Yeah?" I affirmed, getting caught up in the emotion. As a lesbian couple, Robin and Cindy rehearsed their response to the seeming inevitability of this question. Cindy recounts Robin's response:

"She said, 'Well honey you don't have a daddy, you have two mommies. But this is who helped us … Now, Matthew Niedner, he helped Momma and Mummy have a baby. He has a part and Mummy has a part and that makes a baby. You know? Matthew Niedner helped us to make you but you have two mommies.' And then Robin said to him, 'If anyone can be your daddy, who would it be?' 'Cause Robin's thinking, oh if he wants a Daddy we're gonna get him one."

At this Cindy breaks into an infectious laughter and gives me a look that says, 'Come on Stuart, you'll do just about anything for your kids?'

"'If you could have any Daddy who would you want to be your Daddy?' And he says 'Momma'."

"Yeah." And I am caught up in Cindy's joy and laughter.

"So the joke is that I'm his girl-daddy." She says.

"Yeah."

"You know, that is the joke, but that is what he said, he was really little."

"How old was he when he said that?" I ask.

"How old was he? Golly, I'd say three. Yeah."

"Old enough to make an important connection." I stop laughing because I think this is an important point. Cindy stops laughing also.

"Yeah, yeah, he needed us to remind him again why he didn't have a daddy."

I see this as an important conduit into my interests in the gender roles and relations that colour Cindy's and Robin's parenting, from the perspective of the girl-daddy.

"Do you see difference in terms of the way you relate to Wade and how Robin does?" I ask, but I know that this does not get to the specifics about which I am interested, so I add, "you know one of the things that comes up in the early literature on parenting is gender roles. So how do you work through what is traditionally called gender roles between you and Robin?" I fumble, feeling a bit awkward. I recognize this as a leading question, but too late, it is out!

"Mmmmh." Cindy looks at me pensively. Maybe I've fumbled this one right out of the playing-field. Did I just commit the ethnographic faux-pas of leading a respondent into an area that was not part of their lived experience? I back-track.

"Or is that just something you don't think about?"

"No we do have gender roles, we do." Cindy replies. "But it also mixes because I am his mom too. It is a mixed bag. Em, probably we work through the gender roles …"

What a great response! It cuts to the heart of the feminist critique of aggregate studies that posited the existence of definitive gender roles. When Sara Ruddick (1992) challenges Thomas Laqueur's defence of fathering as a right, she argues that we need to understand the work of childcare in order to re-vision fatherhood and motherhood. Her challenge is to critique and unpack the behaviours and responsibilities of parenthood around a set of sexual politics. Jo Foord and Nicky Gregson (1986, 192) indict the notion of gender roles and sexual politics by arguing that the theories restrict understanding the complexities of people's active participation in creating social relations. Roles not only are socially constructed but also are created by people who have choices. Thus, roles are complexly negotiated, modified and changed by the varying degrees to which individuals either accept their existing responsibilities or elect, or are persuaded, to adopt new ones. Gender-role theory is static to the extent that it denies change over space and time.

In the 1990s, Judith Butler's famous notion of gender performativity helped to rethink gender roles and power relations in terms of parents' responsibilities and commitments and, perhaps most importantly, it pointed to the possibility of understanding the ways space authorizes some performances while, at the same time, dismisses others (Aitken 1998, 73). Butler's performativity encompasses the day-to-day performance of gender roles and relations. It cannot be understood "as a singular or deliberate act, but, rather, as a reiterative and citational practice by which discourse produces the effects that it names" (Butler 1993, 2). Butler asserts that all performances are a form of drag, especially the socially mandated displays of patriarchal heterosexuality institutionalized through traditional ideas of fatherhood and motherhood.

Cindy reflects on her negotiated parenting with Robin.

"Yeah, you know it is funny because really Robin is the disciplinarian. I am really soft." She flashes a smile at me and I nod, with a blush.

"You too?" Cindy laughs. "I really am soft and he knows it and he plays it. But the gender role for me really is: I don't like to see everybody get real crazy for no reason. How important is it really? You know, just as the gender role of, you know, I am the one to tackle and stuff. And, you know, my 48-year-old body is not wanting to tackle anymore. You know what I mean? So, I don't know how to answer that."

My question is, as I suspected, moribund. Gender roles say nothing about change and negotiation. And yet Cindy, despite her feeling that her body is failing, enjoys the recognition that comes with performance.

"You know, it comes naturally to me … those gender roles. He thinks I can fix anything. And I am very good at fixing things. You just look at it … And it is not as hard as it seems. You know, but in his mind he thinks that Momma can fix anything so I have to live up to that and so far I have."

"And you are the one he goes fishing with?" I am looking out at the four-foot surf and wondering when we will get out there to cast a few lines.

"Oh yeah! Oh yeah!" Cindy responds, laconically.

"He calls you Momma, what does he call Robin?"

"Mum. Yeah, Mummy and Momma. It is a good deal."

An Althusserian perspective on how power is temporalized would suggest that the gender performances authorized by motherhood and fatherhood, and reproduced in children, are achieved mostly through repetitive performances and the continuous reiteration of ritualized practices. There are, nonetheless, necessary interruptions and productive intervals of discontinuity. Changes in gender performativity are evident when the commitments of family members are redefined, weakened, broken or abandoned as parents' interests change. There is always what Judith Stacey (1990) calls a "structural fragility" to family life and this increases as state and government commitments roll-back responsibilities for child-care – including housing, health and education – to parents. Political and community commitment often is about upholding the myths but not the reality of family living.

Cindy has lived through excruciating family change.

"There are different families today," she notes, "just different families. I mean and there is, look at my family, there was just one mom and no daddy, sometimes there are two daddies; sometimes there is only a daddy; you know there are all kinds of different families."

"When Dad Was Home, He Was With Me"

I want Cindy to tell me about her dad's influence on her parenting.

"Your dad left when you were nine years old," I note. "And you said you know your dad loved you when you were a kid, for the first nine years of your life. How did you know that?"

"How I know that was, I really think, [the] love language he showed to me. It is not words. 'Cause I don't get words; words are just words and they don't mean anything to me, but he would …" Cindy's voice trails off as she remembers. "I remember we built a rabbit hutch together. This big three-storey rabbit hutch out of lumber. He worked at *Dixieline Lumber*. It used to be called *Handyman*. And we had all these Siamese rabbits and we would breed them. And that was the kind of stuff we did 'cause I loved animals and so did he. I see where I got it … He's my dad, it is very clear that our love of animals is the same. But it is by stuff like that. He had a little pool for me, you know, one of these little pools that stand up and he'd put goldfish in it, you know? I had ducks and, you know, we lived in the city: ducks, and rabbits and dogs. He was always in the backyard with me and we'd do this animal thing. Em you know when he came home he would sit in his big chair and I would sit on the arm of his chair and you know we'd always have parakeets and we'd just do this animal thing together. When he was home he was with me." At this Cindy pauses in her memories and feels the emotions behind the stories. Tears glisten in her eyes. "That is it, that is how I knew he would love me."

Cindy switches to a less pleasant recollection of when her father left. She was nine years old and would not see him for the next 30 years.

"Em, but I remember when that last time, he was rambling [drunk] on the door and trying to beat the door down and I remember I always felt so scared and

I remember the trade off. You know, my mom would never press charges against him. I remember this time the cops came and making that decision to, you know," this was a decision that nine-year-old Cindy made with her mom, "we'd better let the cops take him because my fear of him is overriding how much I want him to be here. I think it is a better decision to let him go because I feel scared. Then as I grew up I realized I had all … that just scared the crap out of me as a kid, all that fighting. And so that is where my anger came from, when it was really dissected out of me that's what was under the whole thing, you know, stop fighting you're scaring me, you know, so …"

"So after your dad left your mom looked after you and your sister?" I ask as Cindy pauses. We go on to talk about her anger and how it showed up as a kid at school. She always got into fights, especially with boys. From an early age she watched her mom stand up to her dad's drunken tirades and get knocked down again and again. This was her model of communication.

"I fought boys in school, in elementary school, you know I fought. I did not know how to communicate. It hurts when you get hit by a boy: boys hit really hard, you know?" Cindy laughs that infectious laughter again. "So I had to get all this armour on, this persona to make the boys afraid of me so I didn't have to get in that position of getting hit, though you can't always avoid it. But you know the armour started at a very young age. You know, I suited up and I put on this big tough armour."

Cindy and I talk more about some of our traumatic childhood experiences and then the conversation turns back to her parenting and how her fears show up in her parenting.

Fears of the Father

"I just feel like I am doing everything, just keeping my nose out of the water, barely, and getting some big gulps of water every now and again. You know, and where does my child come in, you know it seems to me that all of a sudden I am seeing my dad." Cindy pauses and sucks in air. The noises from the beach outside diminish, squeezed out of the motor-home like a fist crushing a tube of toothpaste.

"You know, exactly like my dad … he couldn't be there. I can remember on one hand the number of …, the times, the memories that I have with him, you know, and I don't want it to be like that, you know?"

Cindy's anger peals away and she burst into tears. "You know and I think somebody said …," she gulps back the tears, "somebody said one time – I don't know I read it in some book or heard it on Oprah – I don't know what, but somebody had said that there is this statistic that whenever your dad left then that is when you will leave your children," more tears, "and that just plays in the back of my head, and I hope that isn't true, you know, how do you guard against that happening? My dad left when I was nine …"

Another gulp of tears and a huge sigh, “Ah, you know I just feel myself not being present for my son, ’cause everything seems so hard … so, I got to fight for this, you know? And eh, you know, change the books, I don’t know.”

“You don’t have to fight,” I say, “there is a miracle here. Look at what you’ve done. You can change things. You can change the cycle, you can change the model. You know I don’t really talk about past stuff I am doing with other interviewees, but what I am seeing, and a huge trend here is – and I am talking to fathers – is fathers who are trying very hard not to do what their fathers did, and succeeding. So you don’t have to … you, you can break the cycle …”

“I’ve broken it so far”

“Exactly, yeah!”

“But I am not at that nine year point.”

“Right, right.”

“Did he just hold on for nine years, I don’t know!”

“But then you create this nine years as this huge symbolic, you know, area … I’ve heard people share in AA meetings about relapsing and they have this thing about relapsing, say, every two years and sure enough …”

“Right, they create it.”

“And it doesn’t have to be that way. You and I know that we can change.”

“Right.”

“You and I are testimony to the ability to change that cycle. You’re gonna be okay.”

“Yeah, I just hope I have what it takes, you know?

“What does it take?

“Guts, some guts, it takes some courage, it takes courage.”

“Absolutely, you’ve got courage. And you don’t have to do it on your own. That is the key for me.”

“Yeah.” Cindy grabs a tissue and looks me in the eye. “Yeah, yeah that is cool because we kind of re-invent ourselves. It is not even like we re-invent ourselves because we become these people, it is not like we’re becoming the old back-to ourselves, it is like into this whole new person, it is whole new life, just one thing at a time in all of our affairs.”

“You are not your dad,” I say. “You are re-inventing parenting. It is new every day.”

“I have broke the mould. I have absolutely. Yeah.” Cindy’s laugh, once more, is infectious.

Affective Fathering

It seems to me that subjection – of the relations between being and becoming – is an important issue in geography that is exquisitely theorized but rarely looked at empirically.

Theorizing representations was a big issue in geography through the 1990s, when the discipline looked to textual and visual metaphors as an aid for understanding subjects and their relations to landscapes and culture (Barnes and Duncan 1992). Some of the first studies of geography and film criticized the discipline's unwillingness to look beyond the material world to the ways culture is co-created with representational and visual imagery (Aitken and Zonn 1994; Benton 1995). Others criticized this move away from the material conditions of lived experience charging that it was also a move away from saying anything about resistance and change (Smith 1993). At the same time – as I noted in Chapter 4 – feminists geographers moved their concerns with material existence to an explicit consideration of bodies (Callard 1998), and a recognition of their efficacy as a "cultural battlefield" and "the geography closest in" (Simonsen 2000, 7).

Cindy's story points to this cultural battlefield in important, affective, ways.

To augment the differentiated capacities of fathering, representations of fatherhood must be discarded in favour of approaches that foreground the embodied, visceral nature of existence, and encourage fluid affective relations such as those that poured (with the tears) from Cindy. Drawing from some of the ideas in Chapter 4, it seems to me that what is needed is a new way of thinking that affirms the ability of fathers to inhabit bodies and spaces in diverse ways, that disavow the constraints imposed by modes of representation that affirm fathers becoming-the-same as their fathers.

This is not about killing old notions of fatherhood. Rather, it requires avoiding positioning fathers in opposition to mothers, men in opposition to women; it means removing the identities of fathers from fixed time and space, and it means not setting up the actual father (and the emotional work of fathering) in opposition to the ideal father (and discourses of fatherhood). It also requires avoiding the suggestion of a resemblance (or lack thereof) of fathers as men or women.

The Power of Repetitions and Becoming Other

Where does this leave me with regard to repetition and citational practices? According to Deleuze, one failure of representational approaches is their inability to distinguish the difference between repetition and generalities. To the degree that generalities are seen to be the effect of repetition, both become subsumed under the same categorical heading. In a series of thought pieces on masculinity, I focus on repetition and the ways it produced stable and known patriarchal structures that I refer to as fractal geographies (Aitken and Lukinbeal 1997, 1998; Lukinbeal and Aitken 1998; Aitken 2006). I argue that images of masculinity (cf. Figure 4.1) repeatedly represent a seeming bargain made through patriarchy[2] that is a

2 I'll have more to say about Deniz Kandiyotti's (1988) notion of the 'patriarchal bargain' in Chapter 8.

series of hierarchies (power struggles in the home, in the corporate office, and in the oval office) that mirror thoroughly known social and spatial orders:

> Like fractals within fractals, the texts and subtexts of [these images] engage a patriarchal logic that is knowable [and] is mirrored continuously so that – like watching images in a hall of mirrors – we learn nothing new of its constituency (Lukinbeal and Aitken 1998, 375).

In the same way that viewing fractal patterns of physical features like coastlines does not change with the scale of the representation, so repetition up and down a hierarchy of scales traces a patriarchal logic of representation. Taken this way, repetition is about representing masculinity in a hall of mirrors and the same image recurs again and again into infinity.

But, as Deleuze makes clear, this is not about generalities, and generalities and repetition should remain distinct from one another:

> In its essence, repetition refers to a singular power which differs in kind from generality, even when, in order to appear, it takes advantage of the artificial passage from one order of generality to another (Deleuze 1994, 3, quoted in Lulka 2004).

This fundamental criticism of representation is significant, for it relates directly to my inability to recognize the multiple characters of men. What I recognize are multiple renderings of the same narrative, the same patriarchal logic of representation. The multiplicities of actual fathers are not located within hierarchical fractal geometries, which are about repetition and sameness, but are brought about by different angles of encounter, by fluidity and movement. And they are exhibited by divergent non-hierarchical spatial formations. As Lulka (2004) points out, these flows and transgressions are typical of material existence, and they are the primary means through which representations and identity are broken down and discredited.

The task of this critical theorizing is not the formation of conclusions or generalizations, but rather the pursuit of concepts that are commensurate with the shifting character of the phenomena they investigate. In the face of the hegemonic patriarchal domination of striated space, more materially realistic notions of disorder are welcome as the keys with which to unlock difference. Put another way, the multiplicities of masculinities become irreducible to identity as their dimensionality constantly shifts. As a result of such movements, fatherhood is de-centred and enters into new relations with itself and others. It is no longer the same because the new embodiments that comprise it encounter one another from new angles. These more transient geographies at minimum suggest the flows, resistances, consent and dissent that typify material realities of embodiment and experience, material realities that Euclidean spaces such as fractals abstract and essentially eliminate in the process of repetition.

Cindy finds herself in a new context as a girl-daddy. And as a father-figure, she is faced with the spectre of the abandoned nine-year-old girl. As her only model of fathering, it traces an indelible line in the sand, a striated space that she faces with fear, anger and sadness. Fear for what she may do when Wade is nine, anger at what her father did, and grief for the loss of that love. And she gets to contemplate a different choice, a smoothing, a line of flight that takes her and her son to a different possibility. It is a choice that affords dislocation from a previous spatial frame; it is a choice that opens up a politics of care and responsibility.

PART III
Moving

Chapter 7
Tactics, Strategies and Lines of Flight

> For the journey outwards towards other worlds today also reveals an uncertain journey inwards; an expedition that exposes tears in the maps and a stammer in the languages that we in the West have been accustomed to employ … It is as though I have fallen into a fold in time, stumbled across a sharp punctuation in the narrative, as my presence, which once apparently flowed effortlessly across the map, is brought up short, diverted, disrupted, dispersed (Chambers 1994a, 245).

The last chapter focused on the emotional work of fathering as an affect that names its own content. By so doing, a spatial frame is made and unmade as fathers are made and make themselves. In the chapters that follow I put fathering frames into motion in different ways. Much is made of late about journeying through generations and across cultures, of telling tales that articulate the isolation of being and the anomie of places, that establish transnational existences and the citizen-self, that move toward communication and co-mingling of identities, that adapt and adopt material affects. In this chapter I return to the emotive representations of fathering that I raised in Chapter 4, but here I put these bodies in motion with the hope that a search for the moving father contrives a series of emotions that conflate into an aesthetic of both care and hope. In Chapter 2 I talked about the search for fathering as a story without a beginning and without an end. And yet, even as a metaphor, stories become empty, or at least insipid, without a corresponding notion of a dislocation that opens up the possibility of surprise. With this chapter, then, I continue the arguments that fitfully bind the rest of the book by beginning a sketch of the importance of movement in/with space and through spatiality/subjectivity to the inchoate, affective post-structural, post-gender notion of fathering that ended the previous chapter.

Moving Journeys

In the beginning epigram, Iain Chambers highlights the unsteady foundation upon which is built any kind of representation or spatial framing. He echoes de Certeau's (1984) sentiment that writing about the dynamism of others and their experiences destroys a temporal dynamic with the creation of *un espace propre* as a place upon which to inscribe narratives and stories. By so doing, Chambers suggests that we fall into a "fold in time" and "stumble … across the map." To avoid creating a frame out of *l'espace propre*, de Certeau (1984, xix) proposes the employment of 'trajectories'; he then dismisses those as "flattening out" experience and opts instead, for a distinction between tactics and strategies, which for him brings space and time into an appropriate tension. Tactics and strategies resonate with the Deleuzian notion of lines of flight.

Movement as a line of flight includes not only broad horizontal shifts across space, but also vertical shifts up and down. With movements there is resistance and there are also moments of conformity. Indeed, conformity is an important spatial aspect of Deleuzian theory, for to focus exclusively on resistance is an act of generalization and purification, effectively loosening and making more mobile the connections between being and becoming. Deleuze and Guattari (1988) develop a multitude of concepts to express the propensity of material to mutate, transform, and thus give rise to a plurality of spatial formations. These include the rhizome, but also the line of flight. And as Lulka (1994) notes, it is the intervals, the spaces in between, the lines of flight, and the movements that remain critical because they do not embody transcendent qualities but rather the experience of existence. In each of these, a final destination is not found. My hope in this chapter is to open fathering up to different politics through an exploration of lines of flight and I am helped in this endeavour, once more, by an imaginary that comes from American movies.

In Part II, I tried to construct something that embraces the strangeness of fathering identities while at the same time moving beyond the prison cell of men's bodies and their positioning. With this chapter I want to take those bodies and positions quite literally on the road. Two movies that raise the context of imagined lines of flight in extraordinary ways are Clint Eastwood's *A Perfect World* (1998) and Wim Wender's *Paris, Texas* (1984). The former is a journey across Texas with convict escapee, Butch, and his hijacked 'partner' Phillip. Eastwood uses this foil to highlight the potential becoming of the perfect father. *Paris, Texas* is the journey of Travis, literally out of exile and out of Texas to reunite with his son in California and then to return to Texas in search of the mother.

A Journey Across Texas

Eastwood places *A Perfect World* (1998) in Texas in 1963. Butch (Kevin Costner) and his cell-mate break out of prison and steal a prison guard's car. Their search for a less conspicuous vehicle takes them into a residential neighbourhood early in the morning. Butch's partner smells eggs and bacon and, on going to investigate, interrupts Phillip's single-mother preparing breakfast. Sensing that his cell-mate is up to no good, Butch bursts in on the family scene, into which the son Phillip has just entered. Tension erupts as Butch and his cell-mate square off against each other and the mother backs into the corner, whimpering. Phillip stands in the doorway. In a show of trust, Butch gets Phillip to pick up his 'pistola' and point it at him and say 'bang, bang'. Phillip obliges with some trepidation, but there is also a hint of a smile. Phillip's dad left the family some years before with a promise to return. The mother runs the family through the strict tenets of the Jehovah's Witnesses and, as a consequence, there is little play and frivolity in Phillip's life. Phillip understands that Butch is offering him a playful moment, and the tension dissipates with an important first connection between Phillip and Butch. But not for long.

The scene is interrupted by an elderly neighbour with a shotgun. Butch reacts quickly, taking Phillip hostage and forcing the neighbour to put down his gun. Butch and his cell-mate run; Phillip is hostage during the manhunt that comprises the balance of the film. But the manhunt is a foil for Eastwood's real story, which is a journey in search of the perfect father.

Phillip's lack of a father figure is something that Butch relates to, and this provides common ground between them. In one scene, when Phillip says that his mother told him his father would return, Butch replies that she lied and that he is never coming back. He doesn't come out and say it, but we sense that Butch knows this from his own personal experience. In a later scene, Butch pulls out a postcard from Alaska and tells Phillip that this is where he is heading. The postcard is the only contact Butch has had from his father. Butch is attracted to the wildness and the isolation and the promise of freedom suggested by the mountain scene on the postcard; less inclined to admit he is searching for his father, Butch tells Phillip that these are the reasons for his journey. His line of flight takes him across the state of Texas, but no further.

Along the way, Butch broadens Phillip's horizons by allowing him to experience things that his Jehovah's Witness practising mother would never condone, like drinking soda, wearing a Halloween costume and going trick-or-treating. On a rolling stretch of highway, Butch secures Phillip to the roof of the car so that he can experience a roller-coaster ride as his "God-given American right."

As they journey across Texas, Butch and Phillip create 'a perfect world.' Both are searching for a fathering connection, which they find through playful enactments, candid conversations, and honest companionship. As they drive, Phillip open ups to Butch about his fears and Butch responds with advice that suggests a growing and genuine care for the boy.

Butch and Phillip's journey across Texas is a tender mapping of father/son relations. In an important scene for what I want to say here, Phillip ponders on how long it will take them to get across Texas, and Butch instructs him on how to use part of his finger ("from the joint to the knuckle") to connect the miles on the map (Figure 7.1). Deleuze and Guatarri's (1988 12) admonition to "make a map and not a tracing" is, once more, highlighted. Tracings kill: the tracing of Phillip's finger on the map represents miles across Texas, and the distance to the closing of their journey where Butch is cornered and shot. Their real journey is not escape, but a line of flight towards a perfect world where fathers and sons connect in a meaningful way. In the final scene, shot and bleeding, Butch makes a pact with Phillip's mother that she removes the strict framework surrounding her son's upbringing. On this condition, he lets Phillip go and, as he does so, he opens an opportunity for a police marksman to finish him off.

The mapping that comprises Butch and Phillip's journey across Texas comprises a series of tender moments where each connects with a part of themselves that is

Figure 7.1 Phillip maps their journey across Texas using his finger as a ruler

about longing and the connection foments surprise. If each in different ways is searching for something that is missing in their lives, then they encounter the possibility of filling that void in a series of touching connections: trick-or-treating, simulated roller-coaster rides, stealing a Casper-the-friendly-ghost outfit. The moments are material externalities, and the surprise, for each, is the recognition of what the moment is fulfilling internally. The movie is a movement towards a perfect world in the sense that the evolving relationship is one of openness and surprise. The effect is all the more powerful given the imperious framework of the Jehovah's Witnesses from which Phillip has heretofore derived his experiences. Butch and Phillip's journey is a cartography of affect.

Tender Mappings

In *Cinematic Cartographies*, Tom Conley (2007) looks at a wide range of movies, from French New Wave to classic Westerns, with an eye towards tender mappings. Conley argues that many cinematic representations constitute emotional journeys that are played through maps. These mappings appear as lines of flight and their purpose is rhizomatic in the sense that they repeat the affect. Following the inspiration of Guilliana Bruno (2002), Conley argues that a *carte du pays de tendre* is narrated through movies to the extent that other imperious and mechanistic cartographies are offset.[1]

1 In a monumental reworking of cartography away from its mechanistic roots, Guilliana Bruno (2002) engages the contexts of mapping in architecture, travel, design,

As an example, Conley focuses on Roaul Walsh's *High Sierra* (1947), which ends with Earle (Humphrey Bogart's character in the performance that made him a star) chased by police along I-395 in the Owen's Valley. As a reformed gangster Earle seeks two things: independence and to be alone. These two affects are story-booked for the audience on a map behind the chief-of-police that fades in and out with scenes of police cars and road-blocks. The map details the towns of Independence and Lone Pine (recognizable as real towns in the Owen's Valley). The police presence is irrepressible: Earle's independence is about to be curtailed and he will die alone (Figure 7.2).

If Bruno's and Conley's arguments shift the context of visual geographies (Bruno's purview includes and goes well beyond maps to incorporate architecture, planning and maps) away from patriarchy and imperialism to a consideration of journeys and a tender geographical imagination, then I want to suggest that Eastwood's tender mapping in *A Perfect World* does precisely the same for fathering: that there is an intensification of emotional life that is possible through journeys and spatial images.

A Journey From and To Texas

Based on a play by Sam Shepherd, Wim Wender's *Paris, Texas* (1984) is another tender mapping of a fathering journey. Venerated as a landmark American road movie that journeys through social and spatial landscapes with classic referential iconography and reverential aesthetics, Wenders' movie is the cinematic

---

housing, planning and film. She takes the history of mapping and contextualizes it in the arts, in desire, and in tenderness. Bruno (2002, xi) calls her reworking of cartographic themes against prevailing imperial hegemony a "sentimental geography." As a starting point of her *Atlas of Emotion*, a work that moves in, between and through 17th century cartographies to 20th century films, Bruno evokes Madelaine de Scudéry's map that accompanies the novel *Clélie* (1654). Scudéry's *Carte du pays de Tendre* is a celebrated allegory for the female association of desire with space, and an exemplar of the ways that cartography is inextricably linked with the shaping of female subjectivity (Benjamin 1986). Specifically, the map (and the novel) highlights important passages and mobilities away from lakes of indifference, dangerous seas and *terra incognitae* to favourable villages and towns of tenderness, large hearts, reflection, sympathy and so forth. In the bottom corner of the map that accompanies *Clélie* are four figures: a man, a boy and two women. It is clear that the man and boy are the travellers and the women are the guides. Picking up on Bruno's argument, Conley (2007, 127) suggests that the map in *Clélie* might have been drawn in opposition to contemporaneous military cartographies, inaugurated by neo-Cartesian engineers under kings Henri IV through Louis XIV. These cartographers redrew the defensive lines of France and designed fortified cities in a time when new siege technologies were changing the ways of waging war. Given that the women are guiding the man and the boy, *Clélie* possibly reminded French society of the world of the *salon* and the space that women craft in opposition to the mechanistic world of warfare.

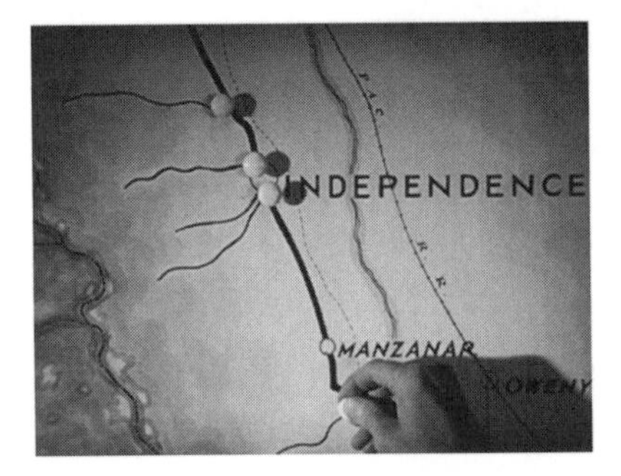

Figure 7.2 The end of Earle's journey is traced on a map

equivalent of Jean Baudrillard's *America* (1988). It highlights popular American places – diner, laundromat, suburb, gas-station, bordello, freeway, canteen, motel – and moves through them with a quirky disinterest or, at best, a sidelong and yet colourful glance. Glimpses of neon motel signs through rain and metronomic windscreen wipers create a persistent nomadic aesthetic.

This is a landscape that passes us by rather than one through which we pass. There is an aching beauty to the places passed and there is resignation in the way we travel as the movie unfolds. There is an unplanned depth, a happenstance, to the landscapes of *Paris, Texas* that engage at a deeply emotional level, and yet there is also a planned depthlessness to the way Wenders guides us through them. *Paris, Texas* is a road movie in the classic sense of the central protagonist, Travis, searching for self through American highways and byways. And unlike other road movies that are searches for independence (*High Sierra*), for community (*Priscilla, Queen of the Desert*, 1994), for lost parental love (*My Own Private Idaho*, 1991) or for an escape from the domination of men (*Thelma and Louise*, 2001), *Paris, Texas* above all else is a quest for and then through fathering. In that this book is a search for fathering as an emotional and geographic practice – the doings that create fathering – and, given that unsteady and amorphous goal, Wenders' opus on American landscapes provides a wonderful parallel.[2]

Out of Amnesia: The Forgotten Father

The journey of Travis in *Paris, Texas* begins with a bird's eye view of the dramatic, desolate mesa and butte landscape of Big Bend National Park in west Texas. The camera glides slowly through steeply weathered, towering, arid slopes to dry washes below as Ry Cooder's guitar twangs a laconic Western lament. The importance of colour in the rest of the movie is highlighted here, in its opening, with a display of sumptuous yellows and oranges against an azure sky. This is not a washed-out landscape and yet it is a classic desert scene that speaks backwards to John Ford and the Utah landscapes he used in *The Searchers* (1956).[3] The camera/bird spots a lone figure walking along a wash with purpose. He is heading somewhere, and yet all the camera/bird sees ahead of him is more desert.[4] Wenders contextualizes

2 By the book's end I intend to distance myself from the image of fathering with which Wenders leaves me, and I do not intend *Paris, Texas* as some exotic leitmotif that structures what I want to say. It is merely a reflection, a staging point if you will, for thinking about tender mappings and the journey towards fathering.

3 The parallels and distinctions between the John Wayne character in *The Searchers* (uncle/perhaps father, lover, lost family and place, loner) and Travis, the Harry Dean Stanton character in *Paris, Texas* (father, lover, lost family and place, loner) are more than coincidental.

4 Dona Harraway (1991) speaks of the bird's eye view as a god-trick. It is, she asserts, the ploy of map-makers, architects, planners, war-mongers and remote sensors of satellite-images to see from above so they may control what is below. Filmmakers use this ploy also, and it is usually called an establishing shot. Remember the bird's eye view of a small twirling figure on a hill in Austria that spins downwards to Julie Andrews and her hills that come alive with the sound of music? Or the small scientific expedition on the arctic icecap that is zoomed in on from high above in *The Day After Tomorrow* (2001). Science's inability to understand let alone stop the earth's reorganization of its weather-systems is highlighted by an opening scene that positions humankind's smallness.

the small figure of Travis from above, wondering in the Texas-desert – perhaps lost or perhaps purposeful and prophet-like. The view is literally that of a bird's eye, we learn, as the scene switches perspective and we watch a red-tail hawk alighting on a rocky crag. Travis stops and turns and looks up, finding the bird. His ruffled black suit, white shirt and black tie suggests that he is out of place and that he has been wondering for a while. His red baseball cap suggests a further incongruity that speaks to madness. Travis holds the gaze of his god and then turns to the horizon. He finishes the water in the plastic jug in his hand and, in classic John Ford style, drops it to the ground and trudges forward, his step with purpose.

What comes next is a beginning search for self, which evolves into Travis' coming to terms with his identity as a father. When he walks out of the desert wilderness he is, perhaps as a nod to Sergio Leone, a man-with-no-name. He seeks water at an out-of-the-way bar near Terlingua, where he collapses. The doctor who treats him is unable to get him to talk, but finds Walter Henderson's business card in his pocket. Walter is Travis' brother and the guardian of his son, Hunter. Walter flies to Texas from Los Angeles, but by the time he gets to Terlingua, Travis has wandered off again.

By chance, he runs into him on a lonely road:

"Travis, don't you recognize me, it's your brother, Walt. What the hell happened to you anyway, you look like 40 miles of rough road."

Travis looks down the road upon which he is heading, but is persuaded to get into Walter's car. It is not until nearly 26 minutes into the movie that Travis talks, and his first word is "Paris."

They are driving back to Los Angeles when Travis shows Hunter a picture.

"This is Paris? It looks like Texas," says Walt.

"It is," says Travis. "Paris, Texas, it is right here on the map." Travis goes on to tell Walt that this is where he wants to go and that he owns the lot in the picture.

"Why the hell did you buy a lot in Paris, Texas?" asks Walt.

"I've forgotten," replies Travis.

Although he has never been there, it turns out that Paris, Texas is where Travis was conceived. It is about beginnings and it is about perceived endings; Paris is symbolic of family and home. It is also a fulcrum around which the movie's journey is hinged, first the road trip from Texas to Los Angeles with his brother and then the lengthier journey from Los Angeles to Houston with his son.

Looking for the Father

In Walter's LA home, it takes time for Travis to gain Hunter's trust. In one scene, which comes after a tumultuous time for Travis in his attempts to reconnect with his son and just after they've watched some old Super-8 home movies that bring them together a little, Walt's Spanish housekeeper asks Travis as he sits flicking though magazines:

"What are you looking for?"

"I'm looking for *the* father," he replies while holding his hand out as if measuring.

"Your father?" She comes to stand behind him with her hand on her hip looking at the magazine.

"No. No. Just a father, any father. What does a father look like?"

"There are many different kinds of fathers, *Señor* Travis."

"Well I, I just need one." He returns to looking at the magazine.

"You think you're gonna find him in there," she taunts.

"Well I don't know where else to look."

"Oh, I see. *You* want to look like a father."

"Yeah!"

The next scene shows Travis looking into a full-length mirror with the housekeeper standing behind him. There is a connection here with the mirrors that Travis stared into when on his journey with Walt from Texas, but this time the angle of the shot is from hip-height, about level with the view of a seven-year-old boy. Travis is trying on a fedora and laughing. The housekeeper asks, also laughing:

"Tell me, do you want to be a rich father?"

"Uh, no!"

"A poor one?"

"No." Travis looks from the mirror to the housekeeper.

"*Pues, ¿cómo pues?*"

"Uh, in-between."

"No, no. There is no in-between, eh? You either have to be a rich father or a poor one." The housekeeper points her finger at Travis' chest.

"Rich," he replies looking back at his image in the mirror.

"*Bueno, un memento. A ver.*" She moves to the closet. Travis steps back from the mirror, still looking at his image.

"*Ahora sí. A vers*, huh?" She says holding up a waist-coat from one of Walt's suits.

"*Qué tal, ¿eh?*" She asks.

"Yeah, *Está bueno*," replies Travis.

"Okay, one thing you must remember: To be a rich father, *Señor* Travis, you must look to the sky and never at the ground, eh? Here!" She puts her fist beneath her chin and pushes it up. Travis puts his head up. "A little higher," she commands. "*Eso*, Mm-hmm. *Ahora camine.*" He walks back and forward across the bedroom.

"*No. Este* – you must walk stiff *Señor* Travis … *Con confianza. ¿Me entiende? Em, con respeto*," she gesticulates.

"Uh, uh, dignity," he blurts.

"*¡Eso! Con dignidad. Ahora.*" She replies with satisfaction. Travis walks across the bedroom and back. "Mm-hmm." She folds her arms and gives a slow appreciative nod. "You got it."

Harry Dean Stanton improvised this scene. It nonetheless speaks to the imperial and hegemonic images that are the frames that the Travis character is journeying beside as he attempts to gain the trust of his son and then reunite with the mother, Jane (Natasha Kinski). It is not until the end of the movie that Travis recognizes the journey and it comes, in part, with a tape-recorded message that he leaves for Hunter:

> Hunter, it's me. I was afraid I'd never be able to say the right words to you in person so I'm trying to do it like this. When I first saw you this time at Walt's I was hoping for all kinds of changes. I was hoping to show you that I was your father. You showed me I was. But the biggest thing I hoped for can't come true. I know that now. You belong together with your mother.

The scene switches from a dimly lit motel room with Travis silhouetted against the window dictating into a tape recorder; cut to Travis listening to the tape while driving and then a final cut to Hunter in a hotel room listening on his own, just prior to Jane showing up:

> You belong together with your mother. It was me that tore you apart and I owe it to you to bring you back together. But I can't stay with you. I could never heal up what happened … That's just the way it is. I can't even hardly remember what happened. It's like a gap. But it left me alone in a way I haven't gotten over. And right now I am afraid. I am afraid of walking away. I am afraid of what I might find. But I am even more afraid of facing this fear. I love you, Hunter. I love you more than my life.

The monologue reflects and refracts a particular line of flight that Travis, the remembered father, is embarking upon.

Making Amends and Moving On

In the longest and most important scene in the entire film, Wenders follows Shepherd's dialogue precisely while creating a now famous (and often copied) visual effect. Travis finds Jane at the Peep Hole Club in Houston. He is in a darkened room with a telephone and she is on the other side of a one-way mirror in a room where she is expected to perform. The performance comes from Travis as he tells Jane a story – his story – and she slowly gains awareness of who is on the other side of the glass.

> I knew these people. These two people. They were in love with each other. The girl was very young, about 17 or 18, I guess. And the guy was quite a bit older. He was kind of raggedy and wild and she was very beautiful, you know? And together they turned everything into an adventure, and she liked that. Even a simple trip to the grocery store was full of adventure. They were always laughing at stupid things. He liked to make her laugh, and they didn't much care for anything else because all they wanted to do was be with each

other. They were always together … They were real happy and he loved her more than anything. He couldn't stand being away from her, uh, the day when he went to work. So he quit just to be home with her. Then he'd get another job when the money ran out. Then he'd quit again. But pretty soon, she started to worry … about money I guess. Not having enough, not knowing when the next check was coming in. So he started to get kind of torn inside. He knew he had to work to support her but he couldn't stand being away from her either. And the more he was away from her the crazier he got, except now he got really crazy. He started imagining all kinds of things. He started thinking that she was seeing other men on the sly. He'd come home from work and accuse her of spending the day with somebody else. He'd yell at her and break things in the trailer. They were living in a trailer park. Anyway he started to drink real bad and he'd stay out late to test her … see if she'd get jealous. He wanted her to get jealous but she didn't. She was worried about him and that got him even madder. Because he thought if she never got jealous she'd never care for him. Jealousy was a sign of her love for him. And then one night, one night she told him she was pregnant. She was about 3 or 4 months pregnant. And he didn't even know. And then suddenly everything changed. He stopped drinking and got a steady job. He was convinced that she loved him because now she was carrying his child. And he was going to dedicate himself to making a home for her. But a funny thing started to happen. He didn't even notice it at first. She started to change. The day the baby was born she began to get irritated with everything around her. She got mad at everything. Even the baby seemed to be an injustice to her. He kept trying to make everything alright for her: buy her things, take her out to dinner once a week. But nothing seemed to satisfy her. For two years, he struggled to pull them back together like when they first met but finally he knew that it was never going to work out so he hit the bottle again but this time, it got mean. This time when he came home at night she wasn't worried about him or jealous. She was just enraged. She accused him of holding her captive – of making her have a baby. She told him that she dreamed about escaping. That was all she dreamed about – escape. She saw herself at night running naked down a highway, running across fields, running down riverbeds. Always running. And always, just when she was about to get away, he'd be there. He would stop her somehow. He would just appear and stop her. And when she told him these dreams, he believed them. He knew she had to be stopped, or she'd leave him forever. So he tied a cowbell to her ankle so he could hear at night if she tried to get out of bed. But she learned how to muffle the bell by stuffing a sock into it and inching her way out of the bed and into the night. He caught her one night when the sock fell off and he heard her trying to run to the highway. He caught her and dragged her back to the trailer and tied her to the stove with his belt. He just left her there and went back to bed and lay there, listening to her scream. Then he listened to his son scream. He was surprised at himself because he didn't feel anything anymore. All he wanted to do was sleep. And for the first time (switch to Travis' face) he wished he were far away lost in a deep, vast country where nobody knew

Figure 7.3 The parent image

> him. Somewhere without language or streets. And he dreamed about this place without knowing its name. And when he woke up he was on fire. There were blue flames burning the sheets of his bed. He ran through the flames toward the only two people he loved but they were gone (switch to Jane's face). His arms were burning, and he threw himself outside and rolled on the wet ground. Then he ran. He never looked back at the fire. He just ran. He ran until the sun came up and he couldn't run any further. Then when the sun went down he ran again. For five days he ran like this until every sign of man had disappeared…

The scene lasts 17 minutes and breaks all the Hollywood rules about rhythm and pace. By the time Travis comes to the end of his monologue, Jane has moved to the mirror. She asks for him to switch on the light on his side and, when she turns off the light on her side of the one way mirror, the audience sees Travis reflection on her face (Figure 7.3). It is a startling image-event. A coalescence, a co-mingling of bodies. It shocks and sways.

Throughout the movie, scenes with mirrors build to this image. When he came out of the desert, Travis stared at his image reflected in bathroom mirrors, motel room mirrors, in the windows of diners. During his amnesia, mirrors were part of his search to find himself, and then later they enabled Travis to find himself as a father. In this final mirror-image, the superimposed faces deconstruct the context of mother and father. Jane's arms held up, simultaneously in abeyance and grasping, signal a tentative reaching out. The image smooths out the contexts of househusband and Mr. Mom, it deconstructs the need for fathering to find form through the maternal. This is precisely what Deleuze and Guatarri (1987) meant by a 'double articulation'. It is also a co-mingling, a leaking of essences and virility. It is a hugely tender moment that offsets – in conjunction with the dialogue – any notion of monstrous parody. This is not the Fountain of Salamis creating a grotesque parody of man and woman, father and mother. She is not traced around him, nor is he mapped onto her in a traditional way. It is a *carte du pays de tendre*. And, with this scene, is fitted an awkward space of fathering, a line of flight. It is freedom from a deterministic form of fathering, but it is a paradoxical freedom. Indeed, what is recovered is something from which Travis never separated. The movie ends with Jane reuniting with Hunter in a Houston hotel room and Travis driving off into a rainy night.

The Awkward Journeys of Fathering

After the journeys of Butch and Travis, as well as Lester in *American Beauty* and Daniel Plainview in *There Will Be Blood*, there is neither an addition nor a subtraction from lived experiences. Lester comes to terms with the beauty of that experience; Plainview is finished by the corruption he started; Butch does not find his father, but rather his fathering; and Travis makes his amends. By its own seamless nature, fathering cannot come as a result of a journey or a mythic quest. It does not come because it is not possible to be apart from it. What intervenes, as in the pathos of Plainview's story and the bathos of Lester's story, perhaps, is the grip of addiction, indulgence and greed. Perhaps it is a focus on self-centred, self-righteous pursuits. This is not to say that the fathers' journeys in *American Beauty*, *There Will be Blood*, *A Perfect World* and *Paris, Texas* are irrelevant to the meaning that Lester, Butch, Travis and Plainview find. The point I want to make with this chapter is that the importance of these movies are not the journeys in and off themselves, but the ways that relatively conventional narrative quests are placed in the contexts of wider scepticism about the roles of fathers and, therefore, wider promise. This is the illness and remedy of these representations: they are simultaneously in and out of the problem and would not be themselves apart from this Janus-headed coin spin.

In many ways, the struggle in all four of these stories is of fathers trying to get back to something that they think they have lost. And yet what they actually do is create new spaces, new bodies and new affects of fathering. If Anglo-American research defines fathers' relationships and involvements with their children as a form of co-parenting that is secondary, or in opposition, to mothering, then the tender mapping that is Figure 7.3 suggests something else, and leads to an opening up of questions about the awkward spaces of fathering that guide the next two chapters. In what ways are men's lives impoverished through social isolation from their families? What issues of gendered power relations and boundary maintenance face fathers and their families? In what ways are relations between domination and subordination in families and the use of space and time in the home and community for family matters relevant to an understanding of the structuring of familial space? These questions circumscribe the problematic social/spatial constructions and frames of fathering, and they also suggest issues on the changing practices of fathers in families, and families in communities.

Chapter 8
Migrant Moves

> Travel, in both its metaphorical and physical reaches, can no longer be considered as something that confirms the premises of our initial departure, and thus concludes in a confirmation, a domestication of the difference and detour, the homecoming (Iain Chambers 1994a, 245).

> Love allows us to enter paradise. Still, many of us wait outside the gates, unable to cross the threshold, unable to leave behind all the stuff we have accumulated that gets in the way of love (bell hooks 2000, 147).

I ended the last chapter with a series of questions about the ways fathers return to reframe and rework the family. The story of Travis in *Paris, Texas* suggests a dance of fathering that flits between patriarchy and the place of mothers and fathers. His journey is one of self-discovery through which Travis realizes that he is afraid of what fathering entails and he is afraid of walking away, but he nonetheless goes to extraordinary lengths to reunite Hunter with his mother. As he sits in his fear, all Travis knows is that through his emotional connection with Hunter he is a father and that he cannot heal what happened. I want to move on now to a story that is not a fictional movie creation and which addresses some of the questions that are left unanswered at the end of the last chapter. These relate to how fathers reunite with the emotional work of fathering, what reparation may entail, and how they return to the space of the family from stories of disillusion. To do so, I follow the story of a Mexican-American father's fall from grace with his first family, his numerous geographic relocations, his attempts at reworking family through a series of marriages, his fight with alcohol dependency and his re-creation of family as an emotional work.

Quixote's dark eyes are the centre of a rugged Latin handsomeness that has got him into trouble on a number of occasions. I remember not long after Quixote and I met I was taking my leave from him in a parking lot and a woman friend rushed after me wanting to know if the sultry gentlemen walking in the other direction was married.

"Even if he is married," she continued, "do you think he's available?"

Quixote does not chase women anymore. That said, he thinks his chosen pseudonym fits well. For most of his life he has chased after the elusive foundation of relationships. This father tilted at the windmill of the family, charged at its audacity and attempted to over-power it and make it his own through self-will. Now, he is embedded in what he describes as "the fellowship" of Alcoholics Anonymous – a large part of his coming community – and from this he derives strength from surrender and vulnerability in the sense bell hooks (1996, 20) means when she argues that although 'surrender' may connote giving up, there are moments when "submission is a gesture of agency and power." Quixote's story is remarkable at a number of different levels: from his beginnings as a migrant labourer from Mexico, to his career as a Marine, to his attaining a PhD in social work and counselling.

I use Quixote's story to highlight fathering as movement – as mobility and disjuncture – told through remembered emotional charges that foment from structural changes and cohere with emotional pushes that comprise a politics of becoming. Said differently, the chapter is about movements (and moments) that precipitate internal shifts through (perhaps beyond, but usually beside) obdurate notions of fatherhood.[1]

A particularly obstinate form of fatherhood that came to the fore in Mexico and the US quite recently is based upon the concept of 'machismo'.

Un Buen Padre es Hombre Pero no Macho

Quixote waves his hand expansively. I join his gaze out towards a broad chaparral-covered valley. Where there are homes the chaparral slopes are replaced by decorative landscaping that meanders between large granite boulders weathered round by centuries of exfoliation. The valley terminates to the south and east in higher mountains. To the west the valley continues, merging with other valleys in a watershed system that empties out into the Pacific Ocean. The rivers in this part of the watershed are too small to sustain water except during the heaviest winter rainfall. I wonder if it is possible to see the ocean on a clear day, given the miles between there and here. We are sitting on the wooden deck of Quixote's suburban home in Southeastern San Diego County, an area known for mountainous terrain, sumptuous homes and panoramic views. The house is on the rim of a canyon and the deck is perched high above the steeply sloped terrain.

Quixote puffs on his cigar and flashes a disarming smile.

"[My wife] found this house and I … I mean I love it," he chortles, "I just love it." Quixote's chortle expands into a laugh and he once more takes in the surrounding panorama with a waving arm.

"Not bad for a little migrant working kid!" he exclaims.

Quixote comes from humble origins in Texas. He remembers moving seasonally around the US with his mother, father and siblings to follow agro-business needs of planting and harvesting. From as young as he can remember, he toiled the fields as part of his father's familial labour force. The six boys and four girls worked with father in the fields and mother provided reproductive support, making meals

1 The chapter draws on the literature of mobile identities and the politics of becoming. Mobile identities are described well through the fluid condition of diaspora, in which a displaced community transplants cultural practices to a new host country with different cultural practices.

Figure 8.1 Quixote's snapshot of a double rainbow on the slope to the west of his home

Photograph used with kind permission.

and clothes, and managing family affairs. The context of fathering that Quixote remembers was elaborated by what he describes as "machismo".

Machismo is a relatively new concept that is often parlayed by US academic and popular publications into a form of sexism and patriarchy. According to Martin Guttmann the term first appeared in Mexico in the 1930s and in the US about a decade later (Guttmann 2007, 372; see also Del Castillo 1993; Guttmann 1996, 2003; Willis 2005). He notes that the semantic roots of 'macho' relate in large part to what is genetically male in plants and animals, and that the term takes on universal cultural meanings in the US that are not evident in a variegated set of meanings found in Latin America. Gloria González-López's (2007, 306) perspective on the culture that surrounds the term in the US is more sanguine; she notes that some immigrant men use the expression *un buen padre es hombre pero no macho* to describe the good father and goes on to suggest that in the US and Mexico today, *los hombres* are "incessantly and creatively redefining gender identities."

This may be the case for Quixote, but a large part of his ideas about, and pride over, family and fathering come from his upbringing and his cultural roots. Quixote's father came to the US as a migrant labourer and lived much of his life with heavily coded gender performances emanating from what Adelaida Del Castillo (1993) describes as a Mexican "covert cultural norm" of patriarchal dominance. What she means by this is a tacit enabling by women so that men may retain a form of male dignity while women control vital aspects of family production and reproduction.

Acts of migration frequently unsettle steadfast assumptions about masculinity and fathering. With this chapter I use parts of Quixote's story to elaborate the constant flux of meanings that complexly compete to define fathering. It is about international migration and the fluid conditions of diasporas, which are paradoxically about change and strong identification with cultural roots. Although Quixote adopts aspects of the model of fathering he saw in his father and moulds them with different forms of nurturing learnt from his mother, his fathering today is also hugely influenced by a myriad of personal contexts – some of which are quite traumatic – that involve dislocation, relocation and reunion.

Part of Quixote's story that is important to note upfront relates to how the affects of mobility are experienced unevenly. Although the circulation of his father was as controlled and contrived as the circulation of capital and commodities, it was nonetheless part of a larger journey that brings Quixote – elsewhere and otherwise – to this moment and this place in Southeast San Diego. How did that hypermobility impact his fathering? How did it change the context of machismo that Quixote's father modelled for him? What were its boundaries and contexts? What were the changes and evolutions?

Elsewhere and Otherwise

Part of the *politics of becoming* is located in a desire for being "elsewhere and otherwise" (Deleuze 1986, 104; Butler 1997, 130). Quixote's dad was a Mexican worker who crossed the border illegally several times in the 1940s and 1950s before deciding to settle in the United States.

I ask Quixote if there was any tension between his Mexican and American heritage, and how living in Southern California changed his views about parenting. Quixote sighed deeply and I realize that he is elsewhere; another time and another place.

"Awe … well, when we came across … when my father came across the border…", he begins before interrupting himself parenthetically.

"He was a naturalized citizen when he passed away." This is an important point for Quixote, whose ideas about nationalism and citizenship are closely linked to a relatively coherent notion of the family as the basis of society. His ideas of family and nation overshadow any patriotism emanating from his service as a career US Marine.

"But, em, he swam across the river a couple of times and met my mom and they decided to get married and stay on this side."

"He swam the Rio Grande?" I interrupted. This was at the height of the Bracero Guestworker Program and the presence of *La Migra* on the border was significant.[2]

2 The Bracero Program was established to attract temporary "guest-workers" as cheap labour for the large bank-owned agricultural businesses that had ousted family farms in the US southwest – particularly California and Texas – in the 1920s and 1930s. The program

Crossing the Rio Grande was always a risky venture for Quixote's dad. Swing describes the beginning of 'Operation Wetback' in June 1954 as a "direct attack … upon the hordes of aliens facing us across the border … Planes were used to locate wetbacks and to direct ground teams working in Jeeps … To discourage re-entry, many of those apprehended were moved far into the interior of Mexico by train and ship" (quoted in Ngai 2004, 155–6). The hardships Quixote's dad experienced crossing the river remain part of family lore.

Quixote's story begins in Loredo, by which time his dad had decided to stay in the US.

"Yeah, down in Loredo, Texas. So, we lived real close to the river."

Quixote laughs, thinking about how his father presided over the family and its economy.

"And, ah, the rule," his laughter fades to a characteristic chortle, "the standing rule in most Mexican households, most first generation households, in the United States, is the father has …" Quixote pauses for effect, "…*total rule*. He has all the power. He has the money. You know?" I nod expectantly, this was getting at what I wanted on changing meanings of fathering.

---

began in 1942, facilitating the movement of 4.6 million workers across the US/Mexican border during its 22 years of operation. A successful applicant for a permit to work in the US would get a contract to work at a particular farm and if caught outside of that farm they would be deported. At the end of the contract – which usually lasted 9–12 months – the worker was required to return home. A text of the time documents the popular view of these temporary migrants:

> Generally speaking, the Latin American migratory worker going into west Texas is regarded as a necessary evil, nothing more nor less than an unavoidable adjunct to the harvest season. Judging by the treatment that has been accorded him in that section of the state, one might assume that he is not a human being at all, but a species of farm implement that comes mysteriously and spontaneously into being coincident with the maturing of the cotton, that requires no upkeep or special consideration during the period of its usefulness, needs no protection from the elements, and when the crop has been harvested, vanishes into the limbo of forgotten things – until the next harvest season rolls around … He has no past, no future, only a brief and anonymous present (Kibbe 1946, 11).

In actuality, farmers who hired workers from the Bracero Program were required to provide housing and meals, and pay a government-prescribed wage. The problem with the program was that a huge volume of illegal migrants – who were prepared to work for less – were attracted across the border with the braceros. In 1954, 'Operation Wetback' was set up to apprehend illegal migrants crossing the Rio Grande. Despite these efforts, the large tide of men crossing the river for seasonal work could not be stayed. The metaphor of an "alarming, ever-increasing, flood tide" of undocumented migrants from Mexico was used by Swing, the 1954 INS Commissioner (quoted in Ngai 2004, 155–6). The Bracero Program was terminated in 1964 after significant lobbying by the agro-businesses that initially advocated the program. There was a collective complaint from agro-business farmers about the advantages of neighbours who used illegal workers and by Union leaders who led an outcry against what they considered abusive living conditions and an undercutting of union wages.

"When he comes home, dinner is ready for him, and when he gets up Mom fixes him breakfast, kisses him, gives him his cup of coffee and his lunch. And she's waiting for him when he comes home. The discipline is all handled by him. You know? He comes home, Mom says 'Quixote did this, Pancho did that.'" His smile broadens with the memory of himself and his brother waiting while his mother explained to his father the two boys' transgression for that day.

"'You need to go talk to them,' she'd say. And we'd be sitting … we'd be forced to sit and wait in the bedroom, waitin' for him, knowing what was coming."

Quixote's smile breaks into a chuckle.

"And then he'd come down and basically he'd just say: 'Your mom tells me you did this.' And depending upon whether he was angry or not – 'cause sometimes he'd get home, he'd be angry about something else going on at work he'd just wallop the hell out of us. You know?"

Total Rule

Did I know? I understood the notion of patriarchy as the rule of the father, but until now I have not considered broader implications.[3] Patriarchy designates structural dominance of women and children by men and, for some, the family is often considered "patriarchy's main paradigm" (Rahman 2007, 469). Alternatively, some contend that although patriarchy may be found in civil society it does not necessarily follow that it is prevalent in private politics (e.g. Elshtain 1990).

The concept of patriarchy re-emerged in academia in the 1970s where it was promoted as a problematic discourse by feminism's second wave (cf. Tongs 1993). It was generally accepted as a concept and practice that precipitated the oppression of women and children, and it continues today in the sense that it signals that many men are advantaged over women and children because of their gender. As noted by Bob Connell (1995), individual men do not necessarily have to support patriarchy to reap its benefits, nor are some men equally able to reap benefits because of a deeper context of marginalization. It is easy to argue that during Quixote's formative years, his dad's context was one of economic marginalization although within his family, others forms of dominance held sway. Connell (1995, 81) is well aware of this kind of dynamic and our difficulty in theorizing around it:

> Though the term is not ideal, I cannot improve on 'marginalization' to refer to relations between the masculinities in dominant and subordinate classes

---

3 In the *Family Fantasies and Community Space* (1998, 102–4), I elaborate the notion of authority in families with the suggestion that mothers and fathers have power over children, but not necessarily in the same ways. I agree with Jean Elshtain (1990) that if "the law of the father" exists in civil society it does not necessarily mean that it exists in private politics. That said, parental authority is special, limited and particular, and the power that emanates from this authority may be abused in ways more insidious and uncontrolled than any other form of power.

> or ethnic groups. Marginalization is always relative to the *authorization* of the hegemonic masculinity of the dominant group … These two types of relationship – hegemony, domination/subordination and complicity on the one hand, marginalization/authorization on the other – provide a framework in which we can analyze specific masculinities. (This is a sparse framework, but social theory should be hardworking.) I emphasize that terms such as 'hegemonic masculinity' and 'marginalized masculinities' name not fixed character types but configurations of practice generated in particular situations in a changing structure of relationships. Any theory of masculinity worth having must give an account of this process of change.

Connell goes on to argue that there is an embodiment of patriarchy in the form of physical violence, or the threat of violence, whether overt or covert, organized or random, individual or collective that underpins the material advantages that men may reap in a particular community or society. From this kind of threat, suggests Deniz Kandiyotti (1988), emanates the 'patriarchal bargain' or 'patriarchal dividend' (see also Whitehead 2007, 467–8).

Kandiyotti argues that the notion of a universal, loving family refuge is pervasive in Western society to the extent that many women and children continue to endure abusive home environments. In *Family Fantasies and Community Space* (1998) I sketch the ways that day-to-day family and gender power relations are constituted within the social imaginary of the nuclear family and the patriarchal bargain. I stress that despite the mythic power of the family, there is no coherent geography of local family contexts and that family studies lack systematic attention to changing valences of space and place. Although the *Family Fantasies* project was primarily an elaboration of the ways that *the* family endures as a mythic and problematic construction within the social imaginary it also points elsewhere, beyond myth to an imaginary that is coherent and focused on everyday practices that revolve around the emotional heart of families.

It is important to note that the concept of a patriarchal bargain is effective only "if we explore gender at a macro level and if we ignore the costs to men of certain forms of masculinity … [it] tends to reinforce, if inadvertently, a notion of gender as ultimately a zero-sum game, one which can only be won by either men or women, not both" (Whitehead 2007, 468). The patriarchal bargain is less compelling as a way of understanding men as fathers when the emotional heart of families is explored. Moreover, when considering the more general evidence of the toll of the patriarchal bargain on men, Susan Fuladi (1999) points out that men are more likely to commit suicide than women, they are more likely to be living alone, and more likely to die younger. Clearly, although mothers and fathers occupy distinct categories of power (as suggested by Elshtain 1990), when considering individual experiences and day-to-day practices of parenting the notion that specific bargains or penalties arise from these categories cannot be assumed. In other words, the notions of patriarchal bargains and penalties, along with the notion of patriarchy itself, are too crude to offer insights into the contexts of familial lived experience.

Movement and Reproduction

Where does this leave Quixote? In what ways does the patriarchal bargain set by his father change with his, and his father's, movement patterns?

Quixote talks about Loredo as a home base for their seasonal migration. By so doing, he moves the context of our conversation from the machismo that characterized his father's construction of household and home to the notion of diaspora, migration and movement as a context of that construction. As part a poor migrant labour family, Quixote's movement with the growing seasons was mandated by economic necessities. Through Quixote, the changes associated with diaspora offer an interesting elaboration of changing meanings of fathering, patriarchy and machismo. The idea of diaspora as identity formation through cultural narrative was introduced by Stuart Hall (1990) in a famous essay entitled "cultural identity and diaspora." He argued that people in diaspora embody the ways that culture is a dynamic, on-going meaning-making set of representations and practices. This dynamic challenges obdurate forms of fatherhood.

"What do you mean 'we'?" I asked. "You were all going with your dad?"

Quixote gives a wry smile over the rim of his coffee cup. "Texas … Loredo, Texas was our home base whenever we got done working the fields in North Dakota, Arizona, Georgia … we went all over the United States."

I asked Quixote when he started travelling.

"Well, eh, we travelled from the time that we were born. I mean *we*." Quixote looks me straight in the eye to make sure the gringo professor gets this last part. "My mom travelled with my dad and we all went as a unit. The more kids you had the more work you were able to do, the more money you were able to earn. So my dad had a very lucrative business doing migrant work. He had enough people coming with him that he'd basically act as a foreman and he'd cut the deal and he would negotiate with the farmer and you know …"

"Here are my labourers," I interrupt, gesticulating.

Quixote laughs.

"Yeah, exactly. Six girls and they can all hoe, you know? And the boys, they can do some hard labour. And I can remember being six or seven … The 50 lb flour sacks my mom used to make tortillas with … well, once they got empty she'd sow a strap on them. Then we were in Georgia picking cotton. We'd be picking cotton, we'd be filling our little bag up dumping it into our dad's bag and then going ahead of him and picking cotton and filling our little bags up. So, everybody contributed to the family. So that is the family unit – that I was talking about earlier – that I grew up with. Everybody worked for the family. You know? Regardless of whether you were working with my father in the fields or you were working somewhere else, when you came home with your paycheck you turned over the paycheck, you know?"

Quixote laughs and picks up his coffee cup, but rather than drink he muses over the lip of the cup.

"My brother, he used to get $10. He turned over a $10 paycheck. He'd get like $5–10." Quixote chortles. "He'd just look at it and go …" Quixote rolls his eyes to emphasize the disparaging look on his brother's face and laughs again. "So, I was five or six years old and working the fields. I think it just about depended developmentally where you were at and how intelligent you were or, I mean, some of that work … a monkey could have done it if you trained them right. You know?" I didn't, but that was okay. "So if you're five or six years old [and] you tell the kid this is how you do it, you know the kid does it." Quixote stops and considers the toil – the labour that a monkey could do if trained – and his expression changes. The emotional heart of that work comes to the surface and his eyes sparkle.

"You know, I was always willing to please my father regardless. You know, I think all of us were. And so we were out there … unless I got tired and then I'd run really far ahead and I'd take a nap." Quixote chortles at that memory but his eyes maintain the sparkle of the previous memory, the one that would do anything to please his father. I push back to that place.

"Always willing to please your dad? Why? What was that based on?" I ask.

"Well, you had … I was competing against ten others for his attention." He laughs and looks at me with an expression that pleads, "surely that is enough." But he continues after a pause.

"First of all, my dad was a well-liked, well-respected man. You know, and all the boys hung with him because that's what we did. You know? But it was, it was just to say …" Quixote searches for the right words. "My dad was not very demonstrative as far as affection goes. Okay?" It is clear that he wants to give me some qualification.

"But when he told you 'you did a good job' or, you know, 'this is my son' and introduced you … that was it … you were in your glory. I was in my glory." Quixote makes a gesture of a father presenting a son to an acquaintance.

Covert Cultural Norms

Quixote is the fourth child in a family of six girls and four boys. He calls the women a "silent majority". They had a lot of power, a quiet forceful pressure on family decision-making. Dad was the bread-winner but Mom and the girls steered family affairs. Respect for the authority of his father kept open disputes to a minimum but, Quixote recalls, many of his father's decisions were overturned through bedroom conversations with mother; the next day father changed his mind. For example, one evening not long after they had moved to Milwaukie, Quixote and his brother came home brimming with the opportunity to go on a school trip to Chicago. Dad came home and gruffly refused, saying it was too expensive. Quixote and his brother went whining to Mom and she said, "ask him again tomorrow morning." They did and he passed on a crisp $50 bill with the admonition not to spend it all.

"We did. We always did."

After a protracted ethnographic study of patriarchy and poor families in Mexico City, Adelaida Del Castillo (1993) describes a context of machismo where men appear to rule families but women preside over the actual decision making. She notes from the context of her work in Mexico City's barrios that women often precipitate and control the communities' economic resources but nonetheless pay lip-service to the seeming rule of men. In Quixote's family, father controlled the production of the family economy but what was done with the money – and particularly how the children were steered towards learning and education – was presided over by Mom. Nonetheless, an imagined imperative – a covert cultural norm – structured the family's gender roles.

"Father wanted the girls to be good wives, because that was the role he understood for women. And the boys, well, they, in particular, had to be good workers." What Quixote offers me is insight into that aspect of his fathering that arises out of the context of manhood into which he struggled. It is an emotional context that coincides with his father's machismo but it also coincides with his mother's values: *un buen padre es hombre pero no macho.*

"And so there was competition between the boys, you know, to get his attention. And so we were all, … the oldest boy had nothing to do with wanting to please him because he was the oldest boy. By virtue, he already had my dad's praises because he's the oldest. So when you're the second from the youngest you have to fight for whatever attention you can get and … and the way you got attention was by being a hard worker."

But Quixote had a drawback to garnering affection though hard work.

"I liked books [and] my dad didn't appreciate that. You know? He used to catch me reading and he'd say: 'Put that frickin' book down and go do something … go pick up rocks, you talk to the old man.' The old man was the owner of the farm. 'Go talk to the old man and see if he's got any work for ya.' And so he was trying to instill in me that the only way you were going to make it in life is through physical labour. You know?"

Quixote stares pointedly at me to make sure I get his next point.

"I proved him wrong." There is a pregnant pause as Quixote holds my gaze and then he waves around at his home and towards his surroundings and laughs.

"Yeah, because I told myself there has got to be an easier way to do this. You know? And I saw the farmer driving around in a big fancy truck. And he wasn't out there busting his ass in the fields. So, I started watching him and what he did."

"Did that create any kind of tension with your father?" I asked, steering the conversation away from Quixote's transformation into the lover of books that eventually earns him a PhD. For now, I am more interested in the gendered power relations that constructed his current fathering practices.

"Oh yeah. Yeah. Especially when I got to be about oh 11 or 12 years old and I started seeing that I could do educationally better than just straight labour and

I started devoting more of my time to my studies. And my mom – who I said encouraged the education – loved me for it. And so, I wanted that reaction because all of a sudden it became a thing about pleasing Mom instead of pleasing Dad and I think that kind of hurt my father also. So there was tension about that."

Clearly I can articulate only a few fragments from Quixote's words to illustrate a much larger evolution. In his cascade of words are buried a travelling story that foments his own fathering experience. Quixote is drawn more and more to books and the comfort of his mother's kitchen and away from his father and his brothers. He chortles as he tells me how things worked in the fields as he matured through adolescence. It is a mature and self-confident chortle:

"Pancho won favour with my father because he was out there working and busting his ass, you know? If it started raining I'm going to have to go in the house." Quixote laughs, remembering the confrontation.

"But Pancho would keep on working, you know, that type of thing. And, eh, so that created a tension between us. And then, you know, my dad wanted drinking buddies. You know, my dad believed that if you work hard, you drink hard. You know, and so … I didn't like the taste of beer when I was growin' up; Pancho liked the taste of beer, you know? And so, I tell everybody, I used to be a chubby little Mexican kid when I was growing up – 12, 13, 14 years old. And so, I don't know, I think my dad was trying to teach me to be strong and be able to defend myself because he, he'd send my older brother Pancho over to hit me and beat me up. You know? And he wanted to see if I could beat him up. And so, when he hit me he'd go back and my dad would say: 'here, you can have his beer.' He wanted … he'd give us a half can of beer and I wouldn't work and that would anger him most and he'd say: 'go beat him up.'" Again the unselfconscious laughter.

"And we'd get into a fight. Well, I learned real quick that, you know, if I fell and started crying then Pancho would go get his beer. And so Pancho would come over … He wouldn't even have to swing at me, I'd fall on the ground and start crying, you know?" We're both laughing now. I see Quixote's wife Margo looking out of the kitchen window to see what the ruckus is about. Quixote continues his story.

"He'd go over and get his beer, I'd go into the house crying and my mom would take care of me. She'd nurture me … this is where I get the tortillas, you know? And she'd butter me up a warm tortilla and I'd sit down and I'd watch her cook."

Quixote loves to cook. He has three step-daughters from his marriage to Margo. It gives him great joy to cook for these girls. This is how he shows his love. Quixote is especially tickled when a birthday comes around for one of his family and he gets to cook for them whatever they want. This is the nurturing that he learnt from his mother. His eyes light up as he recounts the ways he treats his family.

"Yeah – like on birthdays – whatever you want me to cook for you I'll cook. The young one … the little one loves spaghetti so she wants me to teach her how to make the spaghetti sauce and so on. I did that. Eh, the older one, she was into fish and, and meat and stuff like that so I'd cook surf and turf meals for her. My wife

– whatever she wants I just fix. But [the middle one] never allowed me to cook for her until this year. This year for her 18th birthday she wants to have a little party and she wants to have some girls over and she wants me to cook for her."

"Nice! Nice!" I exclaim in response to his enthusiasm. Quixote's expression changes, he wants to talk more about the troubles he's faced with his wife's second daughter. I am enjoying the way his story wafts back and forth between families. Memories, like life, are never experienced linearly.

"Yeah, yeah. So I'd do things but it was like it would be total rejection. I mean we'd be sitting at the dinner table and she'd be sitting there, which is her normal place to sit and she'd be …" Quixote positions his body with his back to me, " … turned away from me and her shoulders turned away from me so she wouldn't have to look at me. And I bought the brunt of that for the entire time."

"How do you feel about that?" I asked.

Movement and Emotional Unfoldings

When Quixote was eight years old, the family moved their home base from Loredo to Milwaukie because of family connections in Wisconsin. He moved seasonally with his family until he was fourteen. Quixote is now in his 50s. Between then and now he was a High School drop out, he joined the Marines, at 19 he married a young woman and had a daughter with her, got divorced, married another women with two step-children, moved to Japan with the Marines and achieved the rank of Gunnery Sergeant, moved to San Diego, got married to Margo and her three daughters, and got a PhD in social psychology. As his story unfolds for me on the deck of his home, his youthful body folds into middle-age through the movements of his words, through his dark flashing eyes and the clouds of cigar smoke, through father-memories, through chortles and grimaces, through raw emotions. The unfolding reveals a corporeality through the sweat of his labours, his failed marriages, his struggle with alcoholism, his maturation as a career Marine, and his agonizing attempts at reconciliation with his second wife and daughter that is mirrored, in part, by his struggles with Margo's middle daughter.

"How do you feel about that?" I asked again as Quixote pauses at the memory of Margo's daughter turning her back to him.

"Ah, it was painful. It was very painful. As a matter of fact I sat down with all three of them. And I told them, I basically told the middle one – I was talking to her … I told her that you know you treat my dog with more respect and kindness than you do me. And I was crying. I was crying. And I told her, 'I have done nothing. I have given you what I would have wanted someone to give my daughter as she was growing up.'" Quixote pauses for a moment. This is an important piece of his fathering. A large part of Quixote focus on his wife's daughters is a reparation for his absence during the first ten years of his biological daughter's life. It is a living amends, if you will, to the daughter Quixote had with his first wife.

"And I said I am not trying to replace your dad but I am here because I am in love with your mom and we're going to be together regardless. But you know I'd really like to have a relationship. It got a little bit better because I think she realized she is able to hurt me. I don't think she knew that she had the power to hurt me until that time when she saw me crying and then after that she still did the stuff but it wasn't as vindictive and it wasn't as stabbing, you know painful as it was before. But there was still the rejection."

Spaces of rejection and abandonment followed (some years later) by reconciliation and a coming together are a large part of Quixote's encounters with fathering. As we talk about his experiences with his step-daughter I am struck by seeming parallels with tensions he experienced with his father as he worked at moving on from his family of origin to create his own family. Quixote was 19 at the time.

"And that created tension with your dad also?" I ask, referring to his plans to get married and leave the family home. He was in the Marine Corps at the time.

"Oh yeah, because he didn't …" Quixote pauses, remembering the sequence of events and the emotions attached to them. "Being out on my own and being away from him … when I decided to get married I called home and asked his permission to get married. And one of the first things he said was 'You still have to contribute to the family' and so I think I was sending like $100 a month home or something like that. And I says, 'but I am going to be married, I'm going to be raising my own family,' and he goes, 'I don't give a shit' and 'this is your family.'"

Creating a Family

"So tell me a little about the evolution of your ideas of fatherhood." This is the first time Quixote has mentioned leaving home and forming his own family, so it seems like as good a time as any to launch into one of my standard interview questions. I continue, prodding. "They came from your family of origin. How have they changed or not as you became a father yourself? As you had a family yourself?"

Quixote's response is immediate and uncompromising. He understands well the patriarchal context of his father's fathering and the struggle to escape to something different.

"Well, my attitude and ideas and beliefs about being a father have done a 180 degrees from where I came from. You know, because I don't believe in using physical force to discipline kids, you know, eh, I don't believe in having children fear you as a way of having power over them so that they'll do what you want them to do. [I believe] that, you know, children need to have their own value system, which hopefully I have some input [into] … [It is a value system that] should prevent them from doing those things that I consider wrong, morally wrong. But at the same time I don't want to scare 'em into it. You know? For, once again, out on their own they're gonna do whatever the heck they want to anyways. So, that is one big difference. It is like I just don't do that. The other thing is, you know, especially being a man in a household of women, ah, even in my first marriage, my daughter … you know?" Quixote's voice trails off with the memory of how he struggled with his first marriage and, as a young man, how he struggled to unshackle himself from his fathers tutelage. And how he failed.

"And what I thought was macho in being the husband and father, it was my first wife. Because I learned in that relationship what I did not want to do. You know? And even although I was only in that relationship for five years, and two years of my daughter living in the house, I look back at that relationship – and some of it had to do with alcoholism – and me being absent. But it was the same way, it was ruling through fear and being very controlling. The ultimate control is that they're not working. You have the money and they don't. You know?"

The morning is passing quicker than I expected. We talk about relationships with women, the importance of showing vulnerability and giving our partners space to feel emotions. Quixote makes some more coffee and brings it out to the deck. Because we are talking about being in touch with what is going on emotionally, I steer the conversation back to his biological daughter, Maria, who is now 32 years old.

"So tell me about your relationship with your daughter then? Over the years …"

"Okay. Ah. Well, my daughter and I have always…," Quixote's face assumes a beatific countenance. There are no words to express the bond with his daughter. "It was instant as soon as she was born. You know? She was my baby."

"Were you there for the birth?"

"Well back in those days they didn't let you into the operating room."

"But you were in the hospital?"

"I was in the hospital. I was there and my wife was terrified because she had a girl. She wanted to have a baby boy for me."

"For you?"

"Ya, for me. And, and I told her: 'its okay, its okay'. You know? And, ah, so the bonding that took place between my daughter and I was that I always slept with her on my chest. It used to terrify my wife because she was afraid I was going to roll over on her. But something happened after I went to sleep with her on my chest it was like, you know, I didn't move, you know, and I just held her. And ah we used to watch cartoons together. She was two months old," he chortles at the memory.

"I taught her how to watch cartoons from the get-go … so I could watch cartoons." His chortle turns into a laugh. "And, ah, we did everything together. When I was home she was my life. Unfortunately when my alcoholism took its toll on me, my relationship with her separated because I separated from my wife. You know, I left my wife and, ah, my daughter. And so I'd see her on the weekends."

Quixote's heavy drinking and carousing with other women were in large part responsible for the separation. And so, his first significant movement as a father was away from his daughter.

"Maria was two years old when you separated. Where were you living at this time?" I ask.

"Wisconsin. Yeah, they assigned me back to Milwaukie because I was a bilingual recruiter and they had a big Mexican population there."

"Are you living on the base?"

"No, I was in the community because there are no Marine Corps bases at Milwaukie. So we're living in the community. A friend of ours that we knew, that I knew, from family rented us their house and so we had a nice little place and, ah, so after Brenda and I separated and divorced I'd pick up my daughter and we'd go out for the weekends and stuff like that and stay with my sisters [so] she gets to have other family around and other nieces and nephews to play with. And then they shipped me out, you know? And so when she was about four years old they shipped me out to Camp Pendleton [in California]." The next significant movement.

"Are you getting on with the mother at this time? Is it an amicable separation?"

"No. It is amicable only when we are in front of the child. The child support, I always had that automatically taken out of my check so I wouldn't even mess with it."

"And you are getting visitation rights at the weekend?"

"Yes, but those weekend rights don't help you when you are halfway across the country and so I used to call. I used to talk to her and I'd send her cards and then shortly after that they sent me to Okinawa, Japan. And so when I got to Okinawa what I used to do is buy myself boxes of cards and I'd send her a card almost every day or every other day. And I'd buy her … you know, just a little note 'Love ya, thinking about ya.' You know? 'Daddy misses you'. And sometimes I'd send a little gift for her you know, but when my alcoholism took effect, you know, it took its toll on me – that was the first year I was there … I was sending letters home, the card home, but I wasn't sending any gifts home because I was drinking like, you know, I had no money.

"And I'd always put it on her mom. Well, you know, I'm giving your mom money so she should be able to buy you a birthday present, you know? It was not a nice thing to do to Mom. So, Mom was always pissed off at me. You know? And she had a right! But I always kept on sending her letters. You know? Little notes so that I always kept that relationship with her."

As Quixote talks of his behaviour when he was drinking heavily in Okinawa, I see in his eyes a clarity and an honesty that comes with time and, for him, for the community of Alcoholics Anonymous. He is clear how much his estrangement from Maria and her mother was in large part a distancing that came as part of his focus on alcohol. It was at this point that Quixote raises his beginning movement from dislocation to reconciliation. It took a long time for any kind of reparation with his ex-wife, but the emotional re-connection with his daughter is automatic. And it hurts.

A Living Amends

"And, em, and then I got sober in '81. So that by the time I got sober in 1981 she was – she was born in '74 – so she was seven years old. Yeah, seven years old and they stationed me over in North Carolina and so I had over a year of sobriety and her mom and I were fighting about child support. I went home to Milwaukie and I picked her up and I was taking her to breakfast and she looked at me and she was really sad and I said 'What's the matter, honey?' and she goes, 'Mom tells me that you're here because you don't love me and you don't want to pay any more child support.' And that tore me up. That tore me up. And, you know, we were sitting across from each other and I moved over and sat next to her and I just told her, I looked down at her and I told her, I says 'I'll never stop loving you. It doesn't matter whether I pay tens of thousands of dollars or whether I pay $50,' I say. 'I am always going to love you.' And the great thing about that – talking about miracles – that summer – this is like in February – that summer her mother had to go in and have some really serious dental work done so she wasn't going to be able to take care of my daughter, Rosey. I call her Rosey although her name is Maria. And so she called me up and she says, 'I have to have this operation. Do you think you might want to have your daughter for the summer?' I go, 'when can I pick her up?' She goes, 'well I have to have the operation in June, she gets out of school in June so you can pick her up in June and bring her back to me in August.' I was there in June and I picked her up and we spent that summer together."

"And you took her down to North Carolina?"

"Ya, so I picked her up. I was driving a Corvette at the time. She loved it. She says, 'take the top down, Daddy.'" Quixote breaks out in his characteristic chortle.

"So, you know, I'd two tops so I took the tops out. And she had a little hat and you know I had my hat and we're just cruising all the way across from Milwaulkie to North Carolina, just, 'wherever you want to eat, if you see a restaurant that you want to stop at just tell me.' I spoiled her, I did. And so … and my daughter had hair. My ex-wife knew that I loved long hair. She never had my daughter's hair cut so here she is. She's seven years old, gonna be eight years old that year and she's got hair to the back of her knees. And so we get to North Carolina and by this time I have a little duplex that I'm buying so she has her own bedroom. I have everything set up for her and I had friends that were married and divorced and they had children so I introduced her to all of them … their kids and stuff like that so she had company while I was at work and always one of my friends' wives would take care of her during the day. Well, one day they showed up at my work and they said Rosey wants to have a talk with you. And so I pulled her into my office and I asked, 'What's the matter,' and she starts to cry and I'm saying, 'stop crying' and 'what's the matter.'" Quixote beings to chortles at this memory.

"And she goes, 'I, I, I know you like long hair.' 'Yeah.' She goes, 'But I wanna get my hair cut.' I said, 'How short?' And she goes, 'short.' And I says, 'Okay give me three good reasons why you should have your hair cut and you can cut your hair.' She sat there, her tears dried up and she just looked at me and she goes, 'It is too hard to manage, I can't wash it ... Mom's got to help me wash it or somebody's got to help me wash it and during the summer I get a heat rash all down my back and butt because my hair is so long and I sweat,' and she gave me a third reason that I can't remember, and I says, 'okay honey, you can get your hair cut, but don't cut it too short.' So, she did, she went with ... my friend's wife. And she says 'I'll take her to the beauticians and she can get her hair cut, but I am not responsible for how short she gets it cut.' And so she went in and she got her hair cut and she came back later on that evening. This was at noon they came and saw me. That evening she brings her back and goes, 'I have nothing to do with this, this was all her decision.' So I said, 'Oh my god, okay, where is she?' She goes, 'she's outside.' So I walked outside and I saw her standing there and she looked really cute and I said, 'Okay where is she?' She goes, 'she's right here' and I said, 'no, where's my daughter.' I am looking over her head, I am looking all around. She says, 'Daddy, I am right here.'" Quixote's chortle peels into laughter.

"Then I looked at her and I said, 'Ah, okay, turn around.' And she had it cut right up to her neck and she looked cute. So I give her a hug and I says, 'It's okay, you're beautiful, I still love you.' And so, that is the way our relationship was. You know, she needs something she calls me up. She helps me get ... at that time she helped me get colour-coordinated with my clothes because I'd take her out and buy her a wardrobe for school and, when school started, and I'd buy myself a pair of pants and a shirt and stuff like that and she goes, 'What is that?' I go, 'well, that's for me.' And she goes, 'oh daddy.' And she'd take me back and return my stuff and get me correctly coloured stuff." He's still laughing.

Tomoko Aoyama (2007, 189) points out that compared to studies of father-son, mother-son or mother-daughter relations, the father-daughter pair has received scant academic attention. He argues that this neglect derives from the weak position of daughters in patriarchal society, in combination with a general neglect of fathering as opposed to mothering. Quoting the work of Owen (1983, 11), he goes on to suggest that most contemporary research focuses on daughters' narratives, which recognizes "a problematic bond, full of ambivalences and longings" with themes that highlight the daughter's loyalty and her need to please the father, as suggested by young Rosey's tentativeness around Quixote. Aoyama points out that most studies follow Owen's groundbreaking work to focus on daughters' reflections and perspectives with very little documentation of the fathers' side of the argument.[4] For Quixote, the summer spent with his daughter is about an emotional drawing together, a reconciliation, a living amends. The ongoing connection is palpable.

"Ah, at 21 she called me and told me she was pregnant and, eh, ah, I am sitting there and I said, 'how the hell did that happen?' She goes 'Well Daddy, don't you know?'" Quixote chortles. "And then, 'This is not a time to be smart with me,' you know and she started to cry and I said ... and we talked a little bit more and she

---

4 An exception to this bias is Vitoria Secunda's (1992) volume on women and their fathers, which gives voice to fathers in the first of its four parts. It is nonetheless geared towards women understanding the impact of "the first man in your life."

said, 'I gotta go,' and I says, 'okay, fine I'll talk to you later,' and as soon as I hung up I felt so bad and this isn't the way it should land."

At this point, I get a good example of Quixote's *coming community* as it is contextualized through his relations with, and commitment to, a recovery program. It begins with an ethical point, and Quixote's recognition that something is not right.

"This isn't what I wanted and so I called my [AA] sponsor and my sponsor says, 'Yeah, you were an asshole.' He says, 'Let's meet for coffee.' So we met for coffee and we talked and I told him and he says, 'Okay, so why are you angry?' And I said, 'Well, she got pregnant.' He goes, 'Naw, that's not it. You're not angry for that … that's something else. What is going on?' And I told him, 'I'm afraid for her … a single parent, I know what her mom went through, you know? I know what a couple of my sisters went through and I don't want to see her go through that.' He goes, 'well, call her up and tell her that.' You know? So I called her up the next day and I apologized for talking the way I did and I told her, 'I am just scared for you. I know how hard it is,' and I said, 'you saw your mom and you lived that life.' And she goes, 'yeah, but it is going to be different because I have you in my life now and we have Mom.' And I said, 'okay, whatever you decide I'll support what you are going to do.' And nine months later I flew to welcome the baby in."

Truth is Revealed Only by Giving Space or Giving Place to Non-Truth

Giorgio Agamben (1993) talks of the coming community as emergent, as 'taking place'. Quixote's community with his recovery, his friends in AA, and his AA sponsor foments in times of crisis to distil, in the case of his daughter's pregnancy, an ethically appropriate response. Agamben (1993, IV/13) notes that:

> Ethics begins only when the good is revealed to consist in nothing other that a grasping of evil and when the authentic and the proper have no other content than the inauthentic and the improper. This is the meaning of the ancient philosophical adage according to which 'veritas patefacit se ipsum at falsum.' Truth cannot be shown except by showing the false, which is not, however, cut off and cast aside somewhere else. On the contrary, according to the etymology of the verb *patefacere*, which means 'to open' and is linked to *spatium*, truth is revealed only by giving space or giving a place to non-truth – that is, as taking – place of the false, as an exposure of its own innermost impropriety.

The idea of dichotomous relations and their conflation – good/evil, nature/society, woman/man, mother/father – come together in complex ways in Agamben's configuration. Acceptance of this quirky paradox, of the combinatorial affects of good and evil that reside 'within' each of us as part of and that is also part of the 'without', an "innermost exteriority" as it is referred to by Agamben (1993, IV/15).

Quixote talks to his sponsor, he reconnects with his daughter and, later, he travels to Milwaukie to be part of his grandchild's birth. In these 'doings', he changes.

"I was there when the – not in the hospital when the baby was born but, as soon as … she brought the baby home the next day and I was there. And I did the same thing with my grandson that I did with her. I have pictures with him sleeping on my chest and watching cartoons with me at two days old."

A smile expands into Quixote's chortle.

"So that little guy and I have a strong bond also. But, you know, if it hadn't been for this program, AA – if it hadn't been for sponsors and people talking to me about how you should be a father, you know, and being really frank with me about, you know: you screwed up, you need to make this right, you need to take account for your actions and you need to be able to say that to them without pointing the finger." Quixote's *coming community.*

"And, so that, you know, that is what I do today, you know … I, I, I was going to say I lead by example … I father by example, you know? But I get that from the men in AA. You know? Ever since I started getting sober I'd see these guys talking about their children. I'd see them going through their trials and tribulations and … I see guys that lost their babies at birth, you know? [I listen to] how they … talk about their baby-brother dying. You know? And it is an awesome experience to witness these guys give testimony about what is going on in their life and how they are handling it. And so that is the only thing that gives me strength out here. You know, kind of like [what] Jim B. [said when I asked him how to deal with my daughter]: 'All you can do is love 'em.' I didn't know that."

Quixote laughs but I see in his eyes the depth of his integrity beneath the humour. This is about his dad and lessons of machismo, and it is about his Marine Corps training. And it is about the bell hooks epigram that begins this chapter.

"I never knew that because my mind goes to revenge. My mind goes to the way I was raised. You don't want to listen to me as a father who is trying to be gentle with you? Heh, I know how to be mean. I know how to be angry. I know how to threaten you and I know how to scare you. I don't want to do that. I don't want to do that. I did that when I was in the Marine Corps. I led by fear and I led by example. You know? And a lot of times that served me well. You know, it doesn't serve me well as a father to do that though. And I've learned that through the examples I get through other men about how they deal with life and their kids."

Los Hombres

Quixote's journey moves in circles that machinate around and through toil, patriarchy and episodes of violence that were part of his upbringing. It hitches onto and then glides through the illness and remedy that was his experience in the Marine Corps. It comprises a series of movements away from and then back to family and relationships. His *coming community* is contextualized in the support of men who are also part of his journey in recovery from alcoholism. Becoming other is, as Massumi

(1992) suggests, a group project. We do not talk about the details of his life when he was struggling with substance abuse, but it is clear that the recovery community is part of Quixote's becoming-other: other than his father; other than an abusive alcoholic; other than a Gunnery Sergeant; other than a father distanced from his children and struggling with the awkward space with which he was faced. Quixote's journey is long, undulating and replete with ironies and contradictions.

The shadows are growing long on the canyon slope below us.

"One last set of questions," I open. "Thinking about the larger context of fatherhood and society since you first became a father, how have you seen American – or Mexican – society change in terms of how fathers are looked at?"

"Aagh." Quixote sighs. "I think one of the things that has happened is that the young, young kids – the children – are allowed to have a lot more freedom because fathers – Mexican fathers, first generation fathers, fathers newly from Mexico – are seeing how other men are parenting their children. How non-Mexicans are parenting their children. And I remember hearing a lot of times, 'My kid ever said that to me I'd back-hand him.' You know? But all of a sudden that doesn't happen. You know? As a matter of fact I've … its been a long time since I heard that behaviour or heard that statement. I've heard it between adults like myself, 'If I ever talked that way to my father he'd beat the shit out of me.' You know? And back then yeah, but today I think kids have a lot more freedom to be themselves by virtue of men – Mexican men – and, and everybody else being more, more flexible and being more receptive to these kids being their own individuals.

You know, I think that affects the discipline of the children, because children when they are allowed to, to go out and express themselves sometimes they take that to an advantage and the boundaries are overstepped. Then the kids are able to do … to act out, you know? And I see that sometimes, the disrespect to the adults, to the parents. But I don't think you can have one without the other. I think you really have to allow children the freedom [so they can] say what they need to say but it has got to be incumbent upon the parent to sit there and setting that boundary that you know, 'I am still your parent so there is a certain amount of respect – whether you believe I earned it, deserved it or not – needs to be there.' I think it has come a long way since I was a kid … for the better."

Quixote's beliefs about contemporary Mexican-American fathering accommodate the contradictions of parental authority and children's freedoms. It is a mix that seems to sit comfortably with him today. I ask him how this transformation affected his relationship with his father to which he responds with the story of his father's death. An important transformative reconnection.

"Well it came together." Quixote looks at me and his eyes are sparkling, but his famous chortle is not heard. A long stillness.

"… just being able to sit together because I told you that my father was not very demonstrative as far as affection goes. But my father's need for love was really brought home to me when he was visiting his sister, his little sister, his younger sister here."

"Here in San Diego?"

"Yeah, here in San Diego and he used to come and visit her and so I got to spend time with him and her. And he expressed jealously one time that – I think it was the second to the last visit before he passed away. He came and I am sitting there and I am …, I said, 'okay, I am going to get some donuts' or something and eh my aunt goes, 'What, no goodbye kiss?' And so I went over and gave her a kiss, a peck on the top of the head … she's really short. So I gave her a hug and a peck on the top of the head and I am walking off and all of a sudden I hear, 'Hey, what about me?' … And that was the first time my father had ever asked for anything from me, and I said, 'Yeah,' and I went over and gave him a big old hug from behind and I kissed the top of his head. I said, 'Your getting a little bit bald up here.'" Quixote laughs, his eyes sparkling.

"And, ah, that was it … And after that we hugged each other and stuff like that. So the last time I saw him alive I was ah, I went to the emergency room where he was at and you know we are sitting there and we're just talking and he just looked at me and he goes, 'I'm tired,' and I says, 'Do you want me to take you back to bed?' And he goes, 'no, I am just tired. It is time for me to go home.' And I knew what he was talking about, he was talking about passing."

Tears well up in Quixote's eyes and his voice falters.

"And I said, 'well you know Mom's been without you for a long time.'" Quixote sighs and his large shoulders shudder. A long pause. He cries openly.

"And I hugged him and I kissed him." Another long pause. I too am crying.

"And that was the last time I saw him awake." Quixote collects himself and wipes his eyes.

"And, ah, when he went into a coma they called me up and I went back … he was laying in bed unconscious to everybody else but I believe he could still hear and he knew what was going on. And I told him that I loved him and that, ah, he was ready to go home and he needed to go. That he didn't need to hang around for anybody."

Quixote and I cry some more.

"And then a couple of days later he passed away."

"Wow."

"Damn." Quixote sniffs

"That's great." I sniff.

"So, so, that is it, you know? Then I of course I had … we had the spreading of his ashes down in Mexico because he wanted to go back to Mexico. So, we didn't ask permission, we just did it. We travelled across the border, in his home state, which is Chihuahua. It was just like this …" Quixote waves his hand around the canyon that rims his home.

"There was no wind and then all of a sudden when we went to spread the ashes the wind just picked up and carried the ashes away. It was a beautiful ceremony. It wasn't a ceremony … It was ceremonious for me. … I had some good friends who came along just for the support. But Margo was there and it was good. I thought it was good. You're home, Dad." Quixote makes a quiet 'sshhh' sound through his lips.

"And the winds came up and carried him away. Beautiful just like this." He waves around his home valley again. "High up in the mountains and the valley just opened up. You could just see for miles. It was beautiful."

"Well that is it." It is late in the afternoon and I am done-in.

"The last thing I want you to do if you would is to choose a pseudonym because I cannot use your real name. Something that maintains the Mexican context if you will?"

"Awe. I dunno, Quixote? Yeah, Don Quixote because I've spent some time tilting at windmills."

"And now you're home."

"And now I am home."

Re-Situating Boundaries

Trinh T. Minh-ha (1994, 9) argues that every voyage involves a re-situation of boundaries. The traveller is both the self that moves – taking with him or her the baggage (some required, some wanted, some best forgotten) that is part of the voyage – and the sojourner across a variegated landscape. With each passing, both landscape and traveller are transformed. The traveller moves with intention and foreboding; events collide and collude; the journey changes and the traveller is transformed; the traveller changes and the journey is transformed. With the movement across neighbourhood, city, state and national boundaries there arises an embodied transformation wherein change occurs at a number of levels. Census data may record a permanent relocation; the traveller may experience hope of a new life, or longing for the return of an old life; there may be dread; there may be excitement; there may be joy; there may be grief. The transformations relate also to larger processes of restructuring: some economic, some cultural, some social. Internal economic pushes propel men and women to move to new areas, new nations and new lives. Guestworker programs attract workers and their cultural baggage to borders and to toil on the other side of the great river. Whatever the embodiment of the individual, the family, the community or the nation, permanency is challenged.

At its most trite, being in motion is part of what it means to be embodied and human, and it is experienced differently for different bodies and consciousnesses. Larry Knopp (2003) points out that our bodies, consciousnesses and creations circulate through space as agents and artefacts of production, consumption and meaning-making. Peter Wollen (1994, 189) argues from an unequivocal position "that diaspora theory (and especially theories of the inmixing of otherness, of hybridity, etc.) should be seen as an advance on essentialist theory …"

With Quixote, I elaborate the connections between fathers and children, movement and being, and I allude to larger economic circulations (of guestworkers, of warriors) from which the notion of fathering is transformed. As Quixote's context of fatherhood changed his idea of fathering mutated, holding on to old, obdurate notions of control through fear that he learned as a child and as a marine. With his alcoholism reaching a crisis, he looked for a different model and found that in his *coming community* and in his 'innermost exteriority', put in place by his father, his mother and his faltering experiences as a new father. From these variegated and mutable circumstance fomented an emotive response that Quixote learned – perhaps from watching his father – that he could ignore only at his own expense and at the expense of those he loved. For Quixote, perhaps this transformation is akin to González-López's (2007) larger transformation of macho fathers to *los hombres*. It is not entirely clear what this means, nor is it my intention to offer clarity. It is nonetheless a breath away from the essentialist machismo characterization of Latin men in general and fathers in particular.

PART IV
Stopping

Chapter 9
Geographic Solutions and Trials by Space

Rex pushes his shoulder-length grey hair out of his eyes. He is in his late 50s and has lived in San Diego most of his life. His life as a father has been turbulent, with a variety of events precipitating a series of moves around the county. These moves came from a variety of sources including job relocation, two divorces, a felony DUI (Drinking Under the Influence) and jail time. For a large part of his employed life, Rex was an electrician for San Diego Gas and Electric, working on the telephone poles and power lines. A few years ago he attained a Master's degree in social work and is now a counsellor in child abuse cases. One of Rex's passions is reading and he enjoys debate on issues of social and labour justice, especially if it enables him to elaborate his views against the current right-wing political climate in the US. His other passion is his two boys and his grandson, who he enjoys talking about at length. Rex talks slowly, with emphasis and care. His sonorous bass fills empty spaces.

"My youngest boy is Manny. Manny … he's a big boy. Simon is physically, about my size, em, except thicker. Manny is, em, … he's got the tall genes in my side of the family. He's six-three, pushing six-three and a half and he's probably around 225 or 230 [pounds]. As time went on, I encouraged the boys to all kinds of things. You know: to work with their hands; to do sports; if they didn't like a sport I'd have them do another sport just so they'd tried it, but I never made it – whether it was academic [or] if it was physical – I wanted them to do the best they could. Simon, the oldest one, grew up and he ended up liking individual sports not team sports. Manny, on the other hand, grew up to love soccer and basketball. He loved the team sports. Both are very good at what they do."

Manny is Rex's biological son and Simon is adopted. Simon was nine months old when Rex started dating the boy's mother, Molly. They got married when Simon was two, and from the beginning Rex took on the role of father-figure.

We are sitting on a small deck at the back of my house that overlooks an alley, which winds down towards a road that leads to the active little centre of my town – some call it the village. I've known Rex for nearly 14 years. We often run into each other (sometimes as much as twice a week) at local AA meetings. There is something about Rex that demands attention. Perhaps it is his laconic bass, perhaps it is his long grey hair; perhaps his steely grey eyes. He is pinning me with those eyes at the moment so that I get the importance of his next point.

"When he was – I want to say three and a half – I adopted him. I adopted him and, em, his biological father didn't want to change his lifestyle from what he was doing. And I know his biological father. Off and on over the course of my first career we worked together every now and then. Em, and I always treated Simon as if he was mine probably because I felt he was, because I was there before he was walking. You know, I went through the whole thing."

Molly and Rex divorced when Simon was 12. This was a hard blow for Simon who is now 38 and only recently began talking to his father again.

"By the time Simon was playing those pick-up games, Molly and I were divorced. We were living over by State College in some condos, Manny and I. Manny came to live with me. Simon would not …" Rex pauses, there is still a lot of pain for him around the years of estrangement from his son.

"Probably by the time Simon was 15 or 16 he quit talking to me because he was so mad at me for drinking and divorcing his mom and it was almost till he was probably 29, pushing 30 before we reconnected. He wanted … he knew I'd be there for him if he needed, but he didn't … He was just angry at me. He didn't see that I had anything to offer because I'd lied to him and his mom. And he'd learned I was unreliable … saying I was going to be somewhere and not doing it. None of the follow-through stuff."

Manny began to get in trouble at school. Looking back at that time, Rex felt that he was not in the place he was supposed to be. To find that place, he tells me, he and Manny moved around San Diego County. First, Rex came to believe that he needed to take the boy away from the city and its negative influences, so he moved to a relatively undeveloped rural area about 15 miles east of San Diego

"My solution to his problem was to 'do a geographic'. And I sold the condo and we moved to Alpine where there was no streets. Because I am going to keep my son from getting in trouble."

Rex got married again and tried settling into a rural lifestyle. It didn't work.

"Manny and I stuck it out in Alpine for a while and then came another geographic … Susan and I got divorced in May … it was final."

Rex is hugely proud of his two boys. A large part of his idea of fathering focuses on what he can teach. We'll get to that later. His fathering practices uncover a deep commitment to doing what it takes to keep his sons safe and to cater for their needs. Sometimes, Rex's actions – like moving to Alpine – are misplaced, but they come from a deep commitment to the needs of his boys. His love and his pride kept Manny with him through the divorces and the jail time, but getting Simon back was a protracted series of painful steps through which he learnt to let go and trust that time would heal his sons' wounds. His relocations – his geographic solutions – did not work. But as we'll see in a moment, a different kind of trial by space, the threat of Manny also leaving, created for Rex an abiding moment in his work as a father.

This chapter is about geographic solutions and trials by space. The last two chapters focused on movements and migrations, with this chapter and the next two I begin to bring the discussion home, to a discussion of stopping and sedentarism. The focus is not so much on the meaning of home, as the processes through which the spaces and spatialities of homes and communities relate to ever evolving notions of fathering. Rex's story is only partially about his series of physical relocations in and around San Diego County because it is also about his attempt to find his place as father of Simon and Manny. Certainly, he went through a series of trials that are manifest in his frequent moves; his 'doing geographics'. But there is another trial, equally spatial, through which Rex was trying to find himself as a father. Each of the fathers' trials I talk about in this book are different, as are the outcomes. There is nonetheless a remarkable similarity between the journeys embarked upon by Rex and Quixote, from the previous chapter, and Stan and Fred in the next chapter. This similarity derives from the contexts of how fathering spaces are produced, how they inculcate power through spatial practices, and how the embodied ideas of fathering are worked through those spaces and power relations.

The Trouble with Space

An important moment for thinking about identity, space and politics was when Henri Lefebvre (1991), in his opening arguments about *The Production of Space*, pointed out that the use of the concept 'space', in popular and in academic discourse, is hugely problematic. He was one of the first contemporary critical theorists to suggest that space cannot be taken for granted and that there are important contexts of power embedded in the ways space is understood. To use terms such as space and place, or to corrupt them with qualifiers such as place identity or belonging, argues Doreen Massey (2005, 17), pushes an imagination "with the implacable force of the patently obvious" and, she notes forcefully, "that is the trouble" with space. To suggest that we identify with places in terms of, say, belonging and comfort and that we move in space from one place to another is, according to Massey, somewhat inane. There is nothing that we can do with that suggestion. In the previous chapter I argue that Quixote's physical moves and his movement of ideas about fatherhood are somehow related, but by so doing I do not imply some form of unique catharsis or connectivity between movement and feelings. Space is not simply a container of activities and there are no simple existential links between places and feelings. Concepts of space, place and scale are much too important to suggest simple Cartesian or phenomenological associations. As I move this chapter and the next two towards a kind of homecoming, I am not suggesting any kind of panacea for the problems of fathering or suggesting a stopping place within which fathers might rest easy. The spatial world is not that simple.

Lefebvre enables me to start thinking about concepts and assumptions of space. His project is a reconciliation between formal abstractions about space, on the one hand, and the physical and social spaces within which we live, on the other. He suggests that although these different kinds of space are interdependent and cannot be pried apart, it is possible, nonetheless, to understand how valences of power work on their interdependent relations. The space of everyday discourse, according to Lefebvre, comprises not only "… the space of common sense, of knowledge (*savoir*), of social practice, of political power …", but also the space of the "commonplaces" such as those found in neighbourhoods, in the town and in the city (Lefebvre, 1991, 25). They comprise everyday neighbourhood geographies that, for the most part, no longer serve (if they ever did) the needs of fathers and children.

> Social space will be revealed in its particularity to the extent that it ceases to be indistinguishable from mental space (as defined by the philosophers and mathematicians) on the one hand, and physical space (as defined by practico-sensory activity and the perception of 'nature') on the other … such a social space is constituted neither by a collection of things or an aggregate of (sensory) data, nor by a void packed like a parcel with various contents, and that it is irreducible to a 'form' imposed upon phenomena, upon things, upon physical materiality (Lefebvre 1991, 27).

Lefebvre's work was influential in turning contemporary theorists such as Massey to consider the ways that we have been taken over by abstractions of space, spatial frames, and spatial representations and spatial metaphors that, ultimately, enervate political action. This kind of spatial framing takes on a reality of its own in the same global processes that produce commodities, money and capital. Moreover, the space thus produced serves as a power base for certain actors: "… in addition to being a means of production it is also a means of control, and hence of domination, of power" (Lefebvre 1991, 26). And for the rest of us, it is a source of disempowerment because it forecloses upon freedom, flexibility and surprise.

Lefebvre brings together perception, symbolism and the social imaginary by focusing upon the ways that their conflicting relations can be synthesized into new codes in a process of *producing space*. He argues that we defamiliarize the idea of 'things in space' in preference for the actual production of space:

> Space is social morphology: it is to lived experience what form itself is to the living organism, and just as intimately bound up with function and structure. To picture space as a 'frame' or container into which nothing can be put unless it is smaller than the recipient, and to imagine this container has no other purpose than to preserve what has been put in it – this is probably the initial error. But is it an error, or is it ideology? The latter, more than likely. If so, who promotes it? Who exploits it? And why and how do they do so? (Lefebvre 1991, 94).

To a large degree, Lefebvre is a spatial structuralist, and, as such, he is criticized for side-stepping post-structural theory in favour of what many consider to be a return to grand theorizing about space. Lefebvre's (1991, 25–6) position against what he calls un-unified theories of space (what we might call post-structural theories of space) is that they are insufficient. He points out that their intent is a practical opposition to what exists in that they attempt to destroy, or at least deconstruct, the spatial codes. As I understand it, Lefebvre feels that these spatial codes have already been destroyed and all that is left are relics (words, images, metaphors) that have little bearing upon lived experience. This is also Neil Smith's (1993, 98) point when he argues that a spatial metaphor has "… an independent existence that discourages as much as it allows fresh political insight." To suggest, for example, that family values are the focal point of US society is to depoliticize the relations of fathers, mothers and children. The term is never used unfavourably, and identities established at other scales – the home, community or nation – are easily rolled into struggles to establish so-called family values. To the extent that the term is used by both the political left and the political right as a form of moral high-ground, it has become value-less because it allows very little political leverage.

The theoretical question of this chapter revolves around the ways certain institutionalized concepts, like family–values, childhood or fatherhood, gain legitimacy and how their usage fails progressive political discourses. It is critical that we not only understand the formative bases of culture, but that we also appreciate the ways in which the concepts are constituted spatially. Lefebvre's (1991, 416) notion of a *trial by space* helps here:

> It is a space, on a world-wide scale, that each idea of 'value' acquires or loses its distinctiveness through confrontation with the other values and ideas it encounters there. Moreover – and what is more important – groups, classes or fractions cannot constitute themselves, or recognize one another, as 'subjects' unless they generate (or produce) a space. Ideas, representations or values that do not succeed in making their mark on space, and thus generating (or producing) an appropriate morphology, will lose all pith and become mere signs, resolve themselves into abstract descriptions, or mutate into fantasies.

A *trial by space* is a legitimizing process that nothing and no-one can avoid, says Lefebvre. For example, 'fatherhood' went through a *trial by space* to achieve legitimacy for a large part of the 19th and 20th century, but that legitimacy is now destabilized due to a number of factors described in Chapters 2 and 3. The notion of 'mothers', on the other hand, gained legitimacy from the mid-20th century onwards and with the women's movement of the 1960s created spaces within which this legitimacy found political form.[1] The important point is that the *trial* – the ensuing legitimization and its destabilization – is marked by action on the part of children, adults, families and communities, and those actions are inherently spatial.

It is not difficult to argue that social relations constitute, form and manage space but, in a very real sense, space is more than an end-product of these processes, it is itself a *process*. The reproduction of space, then, parallels the reproduction of

1 There are a variety of feminist writings that speak to the politics of space. An important early example is Marilyn French's bestseller, *The Women's Room* (1977). Later writers moved from French's focus on homes and suburbia to women 'taking back' spaces of communities (e.g. Roberts 1990) and cities (e.g. Wilson 1991). Other feminists speak to the politics of motherhood as an institution, such as Adrienne Rich's (1977) famous *Of Woman Born: Motherhood as Experience and Institution.*

other forms of identity such as fatherhood and childhood. A trial is a process and a process is a performance. And so, in the same sense that we think of Butler's (1990) identity performativity, we can think also of the performativity of space.

The notion of space having performativity does much more than merely tie it to the actions of people, or suggest that people create and manipulate places and that spatial relations are the end product of that manipulation. Performativity gives space a power because it can be understood as something that reproduces itself. Put another way, performativity "recognizes that 'the subject' is constituted through matrices of [spatial] power/discourse, matrices that are continually reproduced through processes of re-signification, or repetition of hegemonic gendered (racialized, sexualized) discourses" (Nelson 1999, 337). Butler (1990) uses Foucauldian genealogical methods that critically examine and disrupt the production of the spatial frames of psychoanalytic theories that naturalize heterosexual desires and place them outside of the realm of political possibilities. This articulation of bodies and space is important because it enables a move beyond the spatial frames provided by psychological theory, which establish figural spaces of symbolic order (e.g. Freud's Oedipal crisis or Lacan's mirror stage) through which subjectivity emerges from and emanates to material spaces (Callard 1998, 390).

As discussed in Chapter 1, Butler's (1997, 11) notion of "subjection" is "[a] power exerted on a subject, subjection is nevertheless a power assumed by the subject, an assumption that constitutes the instrument of that subject's becoming." How precisely subjects 'become' out of and through – like and in opposition to – the seemingly hegemonic power of dominant discourses is of particular interest because, in an important sense, understanding 'becoming other' requires an opening up of the performativity of space.

Massey (2005, 19) suggests that if space, place and scale – in forms of usage that relate to banal connections – can come to be associated with a habituation and disassociated from their "full insertion into the political" then it is important to understand the ways they can be opened. To do so, Massey embraces Ernesto Laclau's (1990) insistence on the intimate connection between spatial dislocation and the possibility of politics. For Laclau, spatialization – for example, seeing space as a Cartesian container of social activities – is equivalent to hegemonization.[2] Spatial framings – such as Cartesian ideologies[3] – are the product of ideological closure: a picture of the dislocated world as somehow coherent (Massey 2005, 25). If we re-consider space, place and scale as important harbingers of the political, as well as important products of the political, then the lesson from Laclau and Massey

2 It is a form of control, a framing of activities into neat bundles that are then amenable to certain forms of analysis from the plotting of migration patterns over decades, to using economic forecasts to predict future freeway grid-lock, to using design elements such as higher density housing or front porches in the hope of creating a sense of community.

3 The relations between Cartesian logic and identity formation are discussed more fully in Chapter 4.

is that there must be a dislocation, a freeing from cartography as a framing of limited possibilities, to a material spatiality that is fluid and open to surprise. This does not in any way soften the concrete interplays between space, community and identity, but it opens up the post-trial reification of concepts. There is an important aspect of space here that Butler's performativity misses.

A Politics of the Possible

Bodies and space – in particular, the co-production of bodies and space – matter. The lesson from LeFebvre's trial by space for fatherhood is that legitimacy once attained is not immutable. Lise Nelson (1999) argues that while it is important to recognize that foundational norms within enunciations of identity are problematic, and while it is important to understand how the "compulsion to repeat" (Butler 1990, 145) re-signifies dominant discourses in insidious ways, the notion of performativity nonetheless forecloses inquiry into why and how particular identities emerge, their effects in time and space, and the role of subjects in accommodating or resisting fixed subject positions. Indeed, Ian Buchanan (2004, 7–8) argues that Butler's notion of gender is best thought of as a double articulation, and that she "introduces an unnecessary complication (or obfuscation, depending on your perspective) in the form of identification." He goes on to point out that the power of gender as expression "resides in the fact that it does not matter if one chooses not to identify oneself as either 'man' or 'woman', those attributes will still impact on your existence because they are the form for which your body is the content." Nelson argues further that Butler's identity of performativity reproduces a semi-Cartesian dislocation from geography, history, place, context and anti-patriarchal and anti-imperialist struggles. This kind of dislocation is problematic for what it cannot push; it forecloses on a politics of the possible. Left unanswered, for example, are questions of how social and political change occur. How, precisely, do fathers struggle against formative notions of a patriarchal norm that prescribes a 'law of the father'? What spaces facilitate this?

Rex helps me answer these questions.

Spatial Search

"And Manny stayed with you as you moved around?" I ask Rex.

"He came living with me when he was in fifth grade and by then I was living up by State College. [Both boys] used to see me a lot when I lived down at the beach 'cause both of them liked to surf. So, once I left the beach that was the last reason Simon had to come see me. And, you know, I mean …"

"Manny was in fifth grade, Simon would be …" I interrupt. It is a problem I have; my forgetfulness with numbers is countered by interruptions that interfere with the flow of dialogue.

"Manny was in fifth grade so Simon was in high school," replies Rex.

"So when you divorced Molly you moved back to the beach?" I have trouble with places also. Rex fills me in on his geographic solutions.

"First of all I moved to Mira Mesa. Okay, that was quick and handy. It was kind of close. I was working between Kearny Mesa and Escondido a lot and so it was a nice mid-point. I mean the I-15 traffic wasn't asinine yet. You know, you could do it. Within two years of Molly and I divorcing I got three DUIs, one of which was a felony. So I left … I put all my stuff in Mira Mesa in storage and went to see my mom and dad down in Kearny Mesa for about three weeks and then I worked a deal at work so that I took a partial leave of absence – all my vacation – and some excuse-time to do my jail term and then got to come back on because I could work when I went to work-furlough after I got out of jail. Then, after I got out of jail and work-furlough, that is when I went back down to the beach. Em, and the kids were by now up in Ramona. Their mom had sold the house and moved to Ramona. She'd re-married, and he was a very good guy. I have no complaints whatsoever. And he treats her good. Em, him and I actually get along very, very well."

"And Simon and Manny are living with the two of them?" I am still confused with the barrage of spaces and relocations.

"At this time. They are living with them as are his two kids from his first marriage.

"What time period are we talking about here?" Still confused over time.

"Oh, goodness." Rex seems a bit perplexed over timing these placements also. "So that would be … I finally … my last conviction was in 1983. Okay, so, and it was probably around, em, I am going to say '85/'86 when Manny came to stay with me at State College. And he was in fifth grade. You know … he was either right at the end of fifth grade or he was at the beginning because he started middle-school at Main … over at Crawford. And he would have gone to Crawford. Em, Simon only came over to the place once and that was it. And I literally did not see him again until the week before he went into the military when he was pushing 21. There are six years difference between the boys. And then Simon was gone in the military for nine years and after he came back, that is when we started re-connecting. Manny on the other hand … It went … you know he kind of had the best of both worlds in terms of doing what he wanted when I was drinking 'cause I was drinking and I ended up … He was getting a little too friendly on the streets and he knew which bars I hung at, so he would check in with me at the bars." Here is the memory and affect connection once more. Rex is quite clear about Simon visiting him only once before the estrangement that hurt him so much. So, Simon is out of Rex's life for a long time. I wanted to know more about how he connected with his other son.

"Sixth grade?" I ask, questioning at what age Manny was looking for his dad in the bars.

"Sixth and seventh grade. And in eighth grade he got suspended a couple of times. And not realizing … And I, you know, I mean what can I say … I was an alcoholic father and I really wasn't attending to business." This part hurts almost as much as Simon's estrangements, but there is still an important, solid connection between Manny and his dad. Part of this connection, is turns out, is fomented by Manny's feelings over Rex's new wife.

"And what about your new wife?" I wonder where Susan is in all of this.

"Her and Manny didn't like each other and, em, interestingly enough, em, … Molly, the boy's mom, she couldn't stand her. And, eh, boy that one ended up in divorce. But that happened. The divorce happened when I finally got sober. And, eh, so Manny … he was mad being up there stuck in the sticks 'cause there wasn't anything to do up there except you know work. And we had property, we had animals, and of course you know there was … it was up where [the] Viejas [casino] is now but when we bought up there Viejas wasn't there. There was a little market down the road so you had easy access to beer all the time and I drank at home. I didn't drink and drive any more. I didn't want to go to state prison. Em, but he, Manny … I encouraged him at school. I wouldn't work overtime on days when … 'cause living in Alpine he went to Granite Hills High School in El Cajon. So there was one bus down and one bus back. And if you didn't catch the buses you're on your own to get home. So he played basketball, he played Lacrosse, he did a lot of things down in El Cajon and he stayed out of trouble. He stayed out of trouble. And I would always pick him up on the way home, which was dicey sometimes."

Rex watches my puzzled look and decides I am not quite getting where he is going with this discussion of the trials that he is going through as a single-father of a teenage son. There is an aspect that he needs to return to.

"To back up a little, when Manny was eight he came down with type-one diabetes. So you know you always had to worry about his insulin levels and getting his insulin in him. So that had a few scary moments. So you know it was important to have, have people around him be aware. He almost died twice: diabetic comas. Both times [were] before we moved up to Alpine, and I have no idea why. You know. You know, at the time his mom …" Rex pauses and deliberately stops where he is going with this line of thinking. When he starts talking again it is out of a different awareness, an awareness that fomented, perhaps, from the trials by space that had little to do with relocation and everything to do with reconsideration.

"It was all because of me."

"And you are separated from his mom at this time?

"Yeah, he's living with me. Yeah. Yeah. And we've talked about it since. Molly and I became very good friends again. You know, Manny told her I quit drinking and I quite drinking solely because Manny told me he couldn't take it anymore. And, you know, the wife at the time told me if I didn't quit she was going to leave. And I just said, 'yeah, yeah.' You know the usual, I'll play that game 'cause I know she's going to drink and tweak before long. Which she did. But then when Manny told me, and there was something in me that I'd lost the relationship with my oldest boy and I could not – that was what made me quit drinking and learn to become sober. Because I couldn't lose both of them. That was my bottom, if you will. It certainly wasn't jail or fines or ER [emergency room] visits or any of that other stuff."

A huge life change for Rex, a reorientation that related solely to his connections to his boys. The pain of the possibility of losing another son was too much for him, and so he was pushed to find a solution that was not about geographic change but inner change. This was the crux of fathering for Rex and I wanted to know more about the emotional precipice upon which his whole sense of self teetered.

"Tell me about that day Rex? Tell me the story of what happened when Manny gave you that ultimatum?"

"Well, he was in the living room. We were up in Alpine and I was probably out back, you know, taking care of the animals and the fruit trees or vegetable garden or playing with my dog. But I am sure I was drinking. I always drank. You know I mean that is what I did. So I came in the house, you know, for one reason or another. It could have been for anything, you know, to get another beer [or] to see how dinner was coming. I have no idea to this day why it was I came back in the house. And that is when Susan told me that if I didn't quit she was out of there. And eh, like I said, I had told her, 'yeah, yeah, okay I'll quit.' Because I knew that wasn't going to have to stay permanent.

"And Manny just looked over … I'll never forget he had this look on his face like 'Dad, I can't do it anymore.' He just told me, he said, 'Dad, if you don't I can't do it anymore'. And I said, 'Can't do what anymore?' And he says, 'I got to move back with Mom, this is friggin' crazy.' He said, 'I never know what in the world is going to go on around here and you're going to be drunk, she's going to be doing something else. You know, I never know if she's going to go to the grocery store and buy food and come back with another horse.' Which she did several times, em, or another dog or a pig or, you know, whatever, a goat. We had 'em all. Which is okay, I didn't mind the animals. And, em, from him you know, looking back on it I mean this was a self-preservation move because he didn't know if there was going to be damn food in the refrigerator 'cause she'd be coming home with a horse. And I thought she was working, you know, I mean I am a good happy drunk making okay money and come to find out she was probably only working 10–15 hours a week and she was spending most of the time at the casinos. So, you know, it wasn't like she could stop and fill up the refrigerator with food, you know? And I remember telling him, I said point blank, 'Are you serious?' He said, 'Yeah, I'm serious.' I said, 'Okay, then I will.' And I said, 'Well what do you want me to do with all the beer?' and he said … "

At this point Rex start chuckling and his usually serious grey eyes start sparkling.

"He said … 'pour it out.' I said, 'Okay, help me pour it out.' We poured out about a case and a half of talls. I had one bottle of Jim Beam [that] we poured out … I didn't drink Jim Beam when Manny was around, I drank it on the weekends when he was up at his mom's because I promised him when we were back down at State College that I wouldn't drink Jim Beam when he was home anymore 'cause he usually had to put me into bed. And I couldn't do that to him anymore. So we poured it out and then…

"Now I have nothing to drink and I turned into an idiot. Angry. So she left anyway and now Manny and I deal with it. I mean I was on the edge. I was just going *fucking crazy*."

Rex says this slowly, lifting his fist and pumping it up and down for emphasis.

"I don't think I'd started going to AA yet. I started working more and all that. I tried the phoney beers like Sharpe or O'Douls. I think those were the only two out and it took me about a 6-pack to realize that that was stupid. You know, I mean, what's the point? So Manny and I started doing things together and things were kind of going okay. And then I just flipped out. I just could not take not drinking. And so I just flipped out. It happened when I was at work and, and one of my co-workers got hold of the boss and the boss came over and took me down to the DAV [detox for veterans] program and I, you know, I mean I thought I was going to have to go to some 28-day lock-down place to get my head together or whatever. And how was I going to take care of Manny? And she told me she didn't think I needed it because by then I already had like a week and half, two weeks without any alcohol so she said you don't need to go to detox, you're already detoxed. And, eh, they got me into an outpatient program which was really, really good and really, really helpful. But the deal was I had to go there on Monday, Wednesday, Thursday nights and Saturday mornings, four hours each and plus I had to do four AA meetings a week. Well, I'm out in Alpine and that's down in Hillcrest, plus I am working out of Otay."

So much for Rex's geographic solutions.

"So I talked to Susan and she came home to take care of Manny while I was going through this."

"And you guys have split?"

"We've split. We were thinking of maybe getting back together, but no. As soon as I got done with the program she was gone. I filed for divorce. I got sober in October '91 she was gone by January '92. February, she was gone by February. And so it is just Manny and I up there and so I told Manny I said, 'okay, we're staying up here until you graduate.' He was a senior. He was going to be a senior. We had one more year and I said, 'When you graduate we are going back into the city.' And he said, 'okay' and eh, Susan and I got divorced in May of that year, it was final.

"Manny and I stuck it out and we took care of each other and I didn't drink. I went to [AA] meetings and he was okay with that and he was datin' and doing his sports and you know we were really doing okay. Financially we were okay. Everything was okay, just that we weren't down in the city. [But things began changing] and, truthfully, em, I had to get rid of most of the animals, which was my joy of being up in the sticks. That and no neighbours: land, having animals and all that. But there was no way I could do it all [without Susan]. And I was ready to come back down into the city anyway, you know because I still had Manny and I am figuring he's going to be around for a while and I did want him eh …" Rex pauses as his thought patterns slip latterly to some of the trauma that plagued his fathering at the time.

"Wow, it dawns on me my boy's diabetic and I don't want him driving up and down Highway 8 all hours of the day or night coming from school, coming from work, coming from friends or whatever. So, eh, I told Susan, – 'cause both our names are on the thing, the mortgage – that I was going to put the house up for sale. And she said, 'I'll buy you out,' which she sort of did. She took over the house when I moved out. And [Manny] started going to Grossmont, the Junior College. He was working full time."

"Where were you living?" I ask.

"Eh, we were living in Allied Gardens. Em, I found a house to rent over there close to school. You know, easy access to the freeway for me to go to work [and] for him to go to … he worked in Allied Gardens when I was still working in Otay. And eh, and it was okay. We were doing alright and, em, we both trusted each other. I chewed him out every now and again and he chewed me out every now and again if I did something stupid, you know, but, em, you know … both of my boys will tell you today that they learned a lot from me, em, not just what not to do but a lot of what to do and when, and they also learned a lot from their mom and their step-dad. And both of them are good boys."

Finding a Space for Fathering

Rex's story is about movement, relocation and the search for a place for his fathering. The rural idyll does not work because, for Rex, that geographic solution is not one about which he has control. His continued moves to find a better place to raise Manny certainly reflect his commitment – his trial by space – as a father, but it is the emotional trauma brought on by his son's threat of leaving that refracts the loss of Simon years before and highlights for Rex what is truly important.

Over the years and over many cups of coffee, Rex has relived with me that look on Manny's face, an emotional stab to his heart, a look that changed his life.

Now, years later, Rex's boys have moved on and have lives of their own. Manny is married with no kids and Simon is divorced and raising a son. And there is an important material space to the connection that he enjoys with his boys. It brings together his belief in what it means for him to be a father; it is the material practice of a father who wanted to teach. I ask Rex about his connection with Simon and his grandson and get an inclination of a connection that is primarily an emotional bond but is reified in Rex's idea of fatherhood. It is a fathering form that is material, and sedentary.

"Simon and I are very, very close," he tells me, "we have been for eight and a half years or something like that and we talk all the time and at least once a week we talk on the phone and I talk to my grandson. Em, he has no problem at all with me being with his son. Em, and Simon has told me many, many things in the last five years when we really made a point to talk every week of different things he'd learnt from me while he was kid while I was still around that are with him today and some of them just blow me away. You know he used to sit there and look at

Figure 9.1 "I would give him free reign to any of my books"

my books and he'd say, 'Dad, I am just in awe that you've read all those books and that you can always find an answer for me.'"

Rex tells me a story about his fathering finding form. It happened when he lived with Molly and Simon was in grade school. Rex thinks of himself as an academic rebel, and he is proud that his sons also question the system.

"You know, when [Simon] was in grade school I used to, you know when they started to do book reports you know and the teachers in Poway – they were good people and all that but sometimes they did things. You know they would, they would have this preferred reading list that they wanted the book reports done on or a book of your choice at the bottom. Well, I would give him free reign to any of my books and, eh, and he was in trouble all the time because he'd never do book reports from the school district reading list and they were not topics that, em, em, the school district would prefer students reading. His very first book report was a book called *The Student as Nigger* … caused an uproar. And I defended him to the hilt. He had to get his 'A' because it was an 'A' report, you know? And they told me to run my books through them and I told them, I told them when they were in charge of the thought-police I would consult with them. And … buzz off, he'll read whatever he wants to read. And Simon has a very, em, em, a very good mind. He is very, very smart. He'll be graduated from college finally, em, this coming June with a 4.0 and he's already been accepted into, em, graduate school for an MBA at Nebraska. And, he's come this far with 9 years in the army and raising a kid and, em, just pluggin' away."

"You must be proud," I say.

"I am. And, you know, one of the things that I regret with Simon was there was that huge period of 15 years of disconnect – except if I heard something from his mom that was going on. And, em, he married and had one kid – Norm – interestingly he got married to a woman in the army who had a nine-month-old kid and he raised that kid. Interestingly!"

What goes around, comes around, thinks Rex.

"Him and I talk about that now and he goes, 'Yeah, why would I? … It was just the way it happened and I didn't think there was anything weird about it.' And I said, 'you were nine months old when I met your mom.' And he goes, 'yes, she told me that later on.' 'Kind of strange son,' I said and he says, 'Strange what you can learn is a normal thing to do.' And the kids ended up, it wasn't his biological kid but Norm who is now ten, that is the one he is raising.

"So, the army sent him back to I think it is Fort Hood or Fort Bliss down there near El Paso, and promptly told him that he was going to have to do a two-year tour in Korea, and he just went ballistic. He said, 'you told me I'd been overseas too long, well guess what, I am not going!' And he got out of the army after nine and a half years. He went to work in LA at the Federal Penitentiary 'cause if he continued working for the feds his time in the army would count towards a pension. And she decided – his wife – she needed to go back to Nebraska where she was from, she couldn't deal with LA and all that stuff so she took the kids and left. And I heard something was going on and I called up there and he was all humming and hawing and I said, 'okay,' and this is like four o'clock on a Thursday afternoon or something like that and he wouldn't come out and say it so I just said, 'I'll talk to you later,' and, eh … he lived in this apartment complex on the edge of East LA and, eh, so I just filled up my truck and drove over there. It was in a security complex and I jumped the fence and got in and went and knocked on his door and asked, 'Okay, what's happening?' Well, she was leaving."

"And this was when you and Simon started talking again?"

"Yeah, we started talking. It was at the very beginning of it. So they told me what was going on. She was leaving the next day, taking the kids with her and he was going to be alone and so I took him outside and we talked and I said, 'are you going to be okay?' and he goes, 'Yeah, I gotta figure out how I am going to do it differently.' I said, 'Okay, whose helping you pack up all the furniture and stuff.' He said, 'I got nobody.' I said, 'I'll be back up tomorrow morning.' I drove back up to LA and helped her load the U-haul with her stuff and I sat down and talked with him. And by now I've got some sober time on me and my concern is just helping him get through the deal at the time. Em, and that I think is what really reconnected Simon and I."

"You taking that action?" The doing that is fathering.

"Me just showing up, just taking that action in which … stuff that I would have done back before my alcohol got out of hand with not just my kids but any of my extended family: cousins, aunts, uncles. If they needed help we just showed up and did it."

"Yeah."

"Well, once I got to drinking I can't just show up. Either I don't remember that I'm supposed to be there or I'm incapable of getting there, you know? And, em, ever since then … He ended up moving back there and then, em. Ever since then…"

"Ever since when?"

"To Nebraska … He worked with, he got a job with the DA (District Attorney) and, em, last year was 2006 so 2005 Thanksgiving him and his wife end up getting divorced but they remained very good friends and their focus was on doing the best they could for Norm, which is Simon's biological son. The oldest boy is in a group home. He's … he's got some severe problems and, em, they not only couldn't afford the money to help him get into treatment or whatever, he was not safe to himself or others. So he's in *Boy's Town*, I believe. Has been for a long time.

"So, she died. His wife died totally unexpectedly at Thanksgiving. She had her tonsils taken out and evidently had a, eh, severe allergic reaction to one of the medicines, antibiotics or something, and died. So, he is taking care of Norm and he's doing fine. And we talk at least once a week for an hour. I am going to go back and watch him walk when he gets his diploma. It is the right thing to do."

Teaching Openness

"Tell me some more about right things?" I ask. "You mentioned a couple of things that Simon talks about. What are a couple of things that you do right as a father that you're learning from your kids?" I want to know about what Rex has learnt about his fathering from his sons now that he is sober.

"I encouraged them to be honest." He replies. "Em, I encouraged them to do the right things for the right reasons knowing that that might not always be beneficial to them at the time they were doing it. I encouraged them both to go to college. Manny, em, one of these days he'll finish up. He has one year left at State College. I encouraged them to not go to college to look for an occupation. Go to get an education. Simon is going to get an MBA. What he would much rather get is a Masters in philosophy or history. You know, he wants to teach philosophy or history, which just cracks me up because these are my two loves and I never tried to push my love of philosophy or history or political science or sociology or whatever on them. I wanted them to find their own area that they loved. Em, Manny met his wife over at State at a women's studies course he was taking. You know, he loved those classes. And he certainly wasn't looking for a wife at the time. That's for sure. And I encouraged them to find where … I said, 'I don't care whatever it is you want to do in life to make a living,' I said, 'I want you to find something that you like.' And they said it was like my mantra and I would tell them all that – repeatedly – em, there is nothing worse than having a job you hate because you will get up in the morning already hating the day. Don't get yourself in that bind. And I still tell them that. Now they hear me say it to Norm, my grandson. And he is like in fifth grade and they're going Dad, you know? And then three hours later, you know, they are telling him, 'listen to what your grandpa said about that job.' Norm is going to be okay. He is a very, very good boy. He is very bright. He's intelligent, artistic, likes the arts. He's well rounded. He's got a lot of people back there in Nebraska that really, really love him and he is okay. You know, and he knows he's got 'em out here too and you know, I talk at long length with both the boys."

Rex and I talk some more about how he got into the counselling that he now enjoys so much. He is amazed at how much more closely connected are all parts of his life – and particularly his work and his relationships with his sons and grandson. It was not always this way. His past is one of finding solutions with constant moves around San Diego County, a job as an electrician on top of a telephone pole, and sons who did not trust him. Today, Rex senses a circularity between his core beliefs about what a father should be – a teacher and protector – and his lifestyle. He no longer fights to protect his sons by controlling their behaviours and remaining emotionally aloof. Rather, through letting go and taking care of his, and his family's emotional needs, he finds freedom and connectivity. The context of his fathering is changing and it has been a trial for Rex to come to terms with those changes. Reflecting on these changes he notes:

"When I first became a father, from the nine month mark on with Simon, I think back then fathers were expected to be two things. They were expected to be providers and enforcers, okay? Em, women were in the workplace but that was … the two income family wasn't necessarily the necessity that it is today, for most families. It is kind of like they did and then you had a couple of kids and you stayed home. Basically, you know, Molly and I could have bought any one of our houses, at the beach or the two houses we bought in Poway – qualified, paid for – and lived fine on my income. Wouldn't have had a lot of the extras but we would have done fine. Em, I think fathering today is a different thing. I think fathers today, em, rather than being an enforcer (I hate that word) but it is kind of – and there are still plenty of fathers out there today and plenty of women … you know, use corporeal punishment or whatever on a kid – I just disagree with that but you got to have a means to hold kids accountable and responsible for their actions that are appropriate for their age and size. But I think fathers today are trying their best as a rule to provide. I am not so sure that they are still trying to eh, eh, help guide, help educate. I don't see too many sitting around encouraging their kids to, to dabble in a lot of things.

"I think fathers today are falling into some bizarre trap, em, where it is more important that they be a friend to their kids than a father. And, you know, I don't mean you can't be you know a friend to your kids but your role really is to help prepare them for adulthood and most friends aren't going to help prepare you for your next step because they are just doing whatever you are doing with you, you know, I mean, it is a different thing and I know it is a difficult balance you know I did a lot of things: I used to go surfing and all that with my kids and I taught them how to swim and surf at early, early ages and be all kinds of different things but I think, em, I think part of society today has parents of both genders, but fathers in particular, running scared. You know I think there are fears … a lot of fathers

Figure 9.2 A *where* of Rex's fathering

today are afraid that they are going to get in trouble with the law, with CPS [Child Protective Services], with the school district, with whomever, if they put their foot down. And I don't mean they've got to beat the daylights out of their kid or anything like that but I think they've got them running scared. And I think kids today are being jipped out of parents you know, I really, really do. And part of that is a societal thing where it is all about me, don't you know?" Rex laughs. He knows that he is elaborating part of a patriarchal pattern here.

"You know, I think there are a lot of fathers out there today that are trying to come to terms with their past and they are afraid to face up to that past, whatever that past was. You know, em, and because they are afraid of that somehow the best abilities they have to be a father they can't really do that and then they get frustrated because they are not the fathers that they want to be. It is crazy. And I am not sure it was not the same way a hundred years ago. I am really not. That is my observation over the course of my life. You know, em, my dad did not want me to go off to war. I didn't want Simon to go off to war. I definitely didn't want him to re-enlist to keep going. … And when I felt that sense of relief knowing that Simon was out of Africa for at least then … Craziest thing, the first thing I thought of was now I understand my dad. And, you know, Simon was pretty cool, I mean he, he wouldn't, well he couldn't even tell his own family when he was going to do stuff and where, and when he'd be home. You know, but he was always somewhere or he was always plain clothes undercover in various parts of the mid-East, Europe, North Africa, South Asia. And, eh, he managed to live through it all. And he still wants to be a history and philosophy teacher. When he told me that I gave the phone to my girlfriend, and I did a cart-wheel in my own crippled way. And I am going, 'holy shit' you know?" Rex looks at me and there are tears in his eyes. This is the crux of his fathering; the unexpected that nonetheless ties into his core beliefs about fathering.

"And now he calls me, 'Hey Dad, go to your bookshelf, and here's what I'm looking for,' and I go in there 'cause he knows I've got something on it with a reference list. Okay thanks, you know. And now it is, 'Dad, when are you really going to get a computer so we can email?' And I promised him this year I'd get a computer. Ha!"

The Coming Community

Rex creates a messy and convoluted family community that moves through a series of phases involving his sons, his parents, his ex-wives and, latterly, his grandson. It comes together and separates at various points, it involves movement across, and the creation of, space; it involves surprises and the happenstance along with the planned and some elements of the predictable. Rex is the first to admit that for the most part serendipity and throwntogetherness win out over planning and predictability. What happened for Rex was also a series of repetitions – doing a geographic, for example – that did not address the root of his fathering problems. The failure through repetition suggests an interesting context for the 'trouble with space.' This brings me back to a consideration of the subject and subjection.

If repetition is regulated by a combination of dominant discourses and some capitulation by subjects, then change is possible only with displacement or slippage within the process of repetition either through subjects' becoming-other or through governmental/institutional change. As I discussed in Chapter 6, for Butler (1990, 28) change occurs with the spontaneous emergence of that which is repressed (e.g. non-binary sexuality), but Nelson (1999) points out that this suggests a process or series of convergences that operate spontaneously and autonomously; that is, independent of the subject if not completely outside of it. If I am concerned about space, then I am concerned about questions of agency, knowledge and participation in place and over time. Place, time and relations do not factor into Butler's analysis:

> Her theory of performativity treats any enunciation of identity as necessarily fixed, and any notion of agency as necessarily one that implies an autonomous, masterful subject. In other words, she runs into a bind because she only conceives agency as stemming from an autonomous (pre-discursive) subject (Nelson 1999, 340).

In short, there is a geographic and corporeal embeddedness to consciousness that is not necessarily about spatial enframements. Nelson (1999, 347) argues that this kind of embeddedness suggests that human subjects "do identities in much more complex ways than performativity allows." It is simultaneously a negotiation and a capitulation that is, at times and in certain places, conscious "but it is never *transparent* because it is always inflected by the unconscious, by repressed desire and difference." By extension, human subjects do family values, fathering and community in much more complex ways than performance and representation allows.

There is no doubt that the work Lefebvre and Butler re-focuses concern onto the production of space and the degree to which ideology is inscribed in space and then acted out upon it and with it, but it misses the material and relational nuances of change, flexibility, freedom and surprise that is the opening of political possibilities. It is clear that Western society has been characterized by an abstract spatial framing that is fragmented into *sub*-spaces devoted to the *performance* of specialized, homogeneous activities (Massey 1995, 494), but it is less clear how these may be transformed.

Eve Sedgwick (2003) tries to get at this through affect. First, she distances herself from Butler (and Nelson's critique) by arguing that performativity is most useful as a spatialized mode of thought. The point she makes is that both Derridian deconstruction and gender theory evoke performativity as a form of anti-essentialism. In this sense, performativity is about how language constructs or affects reality rather than merely describing it (see Gibson-Graham 2006). Understanding the production of language, like the production of space, is part of an anti-essentialist project. The same anti-essentialist projects are foregrounded in Foucauldian demonstrations of the productive forces of disciplines and discourses that have claimed to be simple descriptions. That language and space can be productive of reality is a primary focus of anti-essentialist projects. Re-thinking identity, subjectivity and space as fluid, relational and inexhaustible does not foreclose upon the idea of a conscious, thinking – but not necessarily autonomous – subject. Nor does it foreclose upon new and different political possibilities, and it opens up space for Agamben's (1993) 'coming community.'

Chapter 10
Punctured Domesticity

schools, ecstatic
high test scores
single-income families
involved moms
(sometimes overly so)

home, perfect
definitely a yard
not a busy street
a place on a cul-de-sac

but school
the most important thing

(Don, May 2006)

With this chapter comes the stories of three fathers who spend the bulk of their time as primary care-givers and home-makers. Don works at home while looking after his three children, Stan is a single-father, and Fred and his daughter take care of the bulk of the domestic chores while Mom works as a nurse.

According to the 2006 statistics of the US Census Bureau almost 160,000 of the 64 million US fathers are stay-at-home dads. As the book moves towards home, what I want to do with this chapter is look at the diversity within this category of caring, to get away from house-husband and Mr. Mom stereotypes. The chapter is about the patently domestic and it is also about the domicile.

Paul Harrison (2007) highlights a context of home and movement, dwelling and threshold that engages and goes beyond Heideggerian notions of belonging and attachment. Following the work of Levinas, Harrison (2007, 627) argues that the concept of dwelling "is an attempt to think of space neither as a Kantian *a priori* nor as an outcome or an attribute (that is, solely as a 'social construction' or another factor to be factored in or out). Neither a given nor a result."[1] This suggests

1 Harrison (2007, 626) argues that the humanistic work of Martin Heidegger (1923) on dwelling, homecoming and threshold is foundational for understanding performative, poststructural and posthumanist accounts of existence. Elden (2001) and Wylie's (2002) poststructural engagement with space and spatiality, for example, is predicated upon a Heideggerian understanding of dwelling, although the concept is mobilized through exposition rather than through critical analysis.

a politics of residence that relates to fathering activities. The domestic is often conceived as banal – a crucible for mundane practices such as cooking, cleaning and decorating. de Certeau and his colleagues (1998) point out that these practices are far from banal, and are implicated in both material and symbolic terms in a much larger arena of geo-political interest. The idea of domesticity that I want to get to in this chapter is about travels and discoveries, it is about hope and surprises, it is about the raising of children, it is about the work that creates place.

The link with the travelling father of the last three chapters is important to what I want to say here.[2] In the last chapter, Rex's story suggested a miasma of wandering from house to house through San Diego county with no clear connection until he got sober and refocused his life in ways that were healthy for himself and his son. The current chapter follows previous chapters as a movement towards domesticity/domicile and away from patriarchy, or at least some dominant tenants of the rule of the father. Like Chapter 6, part of the motivation is to gender bend in order to highlight an understanding of fathering that is less about gender and more to do with action. And so, by linking fathering to domesticity and domestic acts in ways that complicated John's story in Chapter 1, I hope to raise, reframe and tear down another spatial frame: the home and the hearth.

Mapping the Domestic

Karen Tranberg Hansen (1992, 1) defines the domestic as "a set of ideas that over the course of 19th century western history have associated women with family, domestic values, and home, and took for granted a hierarchical distribution of power favouring men." What is remarkable, she goes on to point out, is how this "set of ideas and practices" have become global as a result of colonial and capital expansion. Although not without local resistance, Hansen argues that domesticity may be understood as a universal phenomenon "just as imprecisely and yet accurately as patriarchy." The domestic is a manifestation of where certain practices of reproduction and production are located. It implies spatial arrangements that serve to regulate conceptions of family and household. "As a primary site at which modernity is manufactured and made manifest," argues Rosemary Marangoly George (1998b, 3), "the domestic serves as a regulative norm that refigures conceptions of the family from a largely temporal organization of kinship into a spatially manifest entity." The domestic bridges the distance between seemingly public issues, she goes on to note, and the private concerns of families.[3]

2 I take part of my lead from Iain Chambers (1994a, 245), who suggests that, "[t]ravel, in both its metaphorical and physical reaches, can no longer be considered as something that confirms the premises of our initial departure, and thus concludes in a confirmation, a domestication of the difference and detour, the homecoming."

3 This pretext, argues David Sibley (2005), enables the powerful to create exclusive home-like places out of spaces considered public (e.g. gated communities, coffee-houses).

In this sense, domestic spaces are closely linked to patriarchal spaces and, as such, are a "ready-made tracing" which "always comes back to the same" and deflects us from a cartography "that is entirely oriented toward an experimentation in contact with the real" (Deleuze and Guatarri 1987, 12–13). In this chapter, I want to open up domestic spaces to Deleuzian scrutiny. I want to create a new cartography, a *carte du temp de tendre* that elaborates the emotional work of fathering in place.

In Chapter 4, I focused on specific Hollywood representations that punctured embodiments of fathering – including the suburban idyll – so that I could say something about emotional mappings. In Chapter 7 I took those mappings on the road to elaborate, in a different way, the emotional work of fathering. In Chapter 8 I took the experiment on a journey with a migrant labourer. In Chapter 9 I began to search for the home of fathering and in this chapter and the next I started to bring that experiment home. By approaching men from these different angles rather than modes of representation and repetitions, it is my hope that a multiplicity of actual fathers are encountered.

I begin the chapter with an ethno-poem derived from my conversation with Don. It suggests a familiar strategy of middle-class home-seeking replete with an enduring focus on good neighbourhood schools and a preference for a house in a cul-de-sac. What I want to do in what follows is raise some contexts of 'fathering in place' through the stories of Fred and Stan. Stan is a single-father, and both Don and Fred have partners who are career-women and less able to spend time at home with their children.

The balance of the chapter focuses on Fred and Stan, but I want to end this section by disrupting the familiar idyll suggested by Don's opening ethnopoem. He and his wife bought the perfect house near the perfect schools, and then

> we moved in here
> and then the twins came in September
> finances are the root of most stress
> (that is my experience anyway)
> selling a house and buying a house and moving here
> and then I lost my job
> I didn't have any marketable skills
> a lot of uncertainty.

Don's experience with domestic bliss and the suburban idyll is punctuated by a larger economic situation beyond his control. The effect of this uncertainty on his fathering is nonetheless remarkable in terms of his becoming other (than his father). Today, Don works out of his home and is the primary care-giver to his son and his twin girls.

> just being able to be there
> to get to spend as much time as I do with all the kids

it is something that I am really proud of
part of that is teaching them to play
soccer, ride bikes, jumping on the trampoline
it doesn't really matter what I am doing
just as long as I am there and I am doing it
my dad was more distant than I was [ever going to turn out to be with my job]
his job required him to work very long hours and be gone a lot
I am very lucky.

A Final Destination

Fred's 'geographic solution' is to stay put: to surround himself with predictability and habituation. He has been married to Irene for 30 years and they have one daughter. Maria was born two months before Fred's 38th birthday. For over 20 years, they have lived in the same house in a rapidly growing community on the periphery of San Diego. When Irene got pregnant, Fred was unemployed: it was during the early 1980s and the unemployment rate in San Diego county was at a record high of 15 per cent. Before Maria was born, Fred was hired on at his old job with the federal government. Irene was working as a nurse. Shift-work took her away from the home early in the mornings and at weekends. Fred spent a lot of time as primary care-giver.

Fred and I are sitting in my living room. I've known Fred for over 14 years and we've enjoyed a number of conversations about his relations with Maria, who was eight years old when we first met. I offer him a cup of tea or coffee, but he refuses. Fred seems quite focused on the business of the morning – the formal interview.

I ask him to tell me about how he got to be the principle caregiver for Maria.

"It was getting to the point where [looking after Maria] was really an important part of my day," he replies.

"When Irene went back to work after three months off, all of a sudden I was the guy who dropped [Maria] off in the morning, picked her up in the afternoon, and, so that was all [quite an] experience, but it was great."

Fred is in his early 60s and is now retired. Maria recently finished her undergraduate degree and is thinking of going back to university for an MA in anthropology. Fred is proud of his daughter's accomplishments. After helping fund Maria's college education he is worried about the family's current financial situation. He tells me that he is not in the financial position that he'd hoped to be in at this time in his life. But that anxiety is superseded by the great joy surrounding his memories of raising Maria.

The Stressed Economy of Childcare

The need for proactive parenting is another outcome of state roll-backs predicated upon neo-liberal policies. Gibson-Graham (2006, 72) note some of the many ways through which the labour of childcare is practised, and how those practices comprise economic communities in specific locales that look less and less like 'capitalism as we know it'. Although it may be argued that childcare is increasingly restructured through state rollbacks, which commodifies the practice in a capital-centric world, Gibson-Graham counter that childcare services are performed in a significant number of non-capitalist sites.[4] They suggest that perhaps what is needed is theorization of a set of economic dynamics that captures the continued generation and regeneration of childcare in which market mechanisms play no dominant role.

Fred's context is the heart of this kind of non-capitalist economy of childcare.

> she was already out of high school by the time
> I quit working all together and stayed home
> Irene and I agreed that I would take care of all of the domestic stuff
> I tried to get the kid to help
> that has been a struggle for me
> you know, I would always get an attitude.
> even after she was out of high school
> I'd get an attitude.
> would you please clean this?
> but I stuck with it
> and then I was flexible.

"The first help I had care-giving was from a friend of ours from our church. She had a couple of young sons and was a stay-at-home mom so I took Maria there one day a week. And then Irene worked weekends too so on the weekends I was on my own. So, you know, I really enjoyed it and it didn't seem to be anything that I couldn't handle, you know? There were a few little blunders here or there but nothing you know – the formula, one time I gave it to her straight instead of mixing it because I didn't realize that that type of formula needed to be mixed. Irene still likes to bring that up. She brought it up recently; I'd forgotten all about it."

Fred's somber demeanor is washed through with a peal of laughter.

"But anyway, other than that it was a joy, it really was, you know, to have the child and I was willing to do whatever I needed to do to be a responsible parent."

This is an important memory.

"It was … it was a real joy to me to have this child and, eh, I just wanted to do everything I could for her, to be a responsible father."

Part of the importance that Fred attributes to care and loving stems from a reaction to his strict upbringing and, in particular, the model of fathering he experienced as a child.

---

4 Gibson-Graham (2006, 72–4) point to alternative and non-capitalist contexts of childcare such as cooperatives, family day-cares, Steiner kindergartens, nonprofits, community-based childcare centres, communal households, extended families with obligatory childcare and so forth.

"I didn't have a father that I …," he pauses and redirects. "You know my father was real controlling and it always had to be his way and he was always also pretty physical so I did not want to be either one of those things. It was just something I knew that I did not want to do, without really thinking about it. [My father] passed away four years before Maria was born. So, anyway, so it was just, you know, watching her grow, taking her to daycare, or ah, you know, it was different the daycares that I took her to, in people's homes, and then picking her up and bringing her home and then when she got into school and – I was the only one with her on Irene's days at work – I was the only one who dropped her off at the 'before and after' care at the school and then I was the one who picked her up. So we had a lot of time together and on the weekends we had a lot of – it was just the two of us usually and so we did a lot of things, we went to the zoo together, you know? She enjoyed doing that and I'd take her to parks early on, as soon as she started walking I'd take her to parks and push her on the swing and have her slide down the slide and all that kind of stuff."

"Were you living in your place in Sunny Oaks?" I ask. Their home of 25 years is a modest California bungalow, nestled in what was once a quiet neighbourhood. The last few years brought constant noise and dust as CALTRANS constructed a nearby freeway extension.

"Yeah, yes that is the only house she's ever lived in."

"And you were working on Point Loma?"

"At that time I was working downtown."

"So you would drop her off at daycare and then head …"

"Yeah, I would usually … Usually I took a bus, eh, a direct commuter bus downtown. My mom lived close by in Fletcher Hills [and that helped, too]. She always enjoyed having Maria come over. She would take care of her, all that kind of stuff. So that worked out well and you know then after she got in school she developed, eh, difficulties." Fred hedges.

Child Advocacy

"It became apparent that she wasn't, you know, she was in some kind of fear where she wouldn't talk in class. In kindergarten, the teacher asked her questions she would just freeze on the spot and wouldn't answer so that was eventually diagnosed as 'elective mutism,' which can be a pretty serious disorder if it is not treated. Fortunately the school recommended us, or referred us to a counsellor who specialized in children. She had been a teacher who had decided to get into counselling and she was willing enough to work with us, she jumped through all the hoops with the insurance because the counsellor wasn't covered under our insurance. And I think that is, we're both convinced that – Irene and I are both convinced that that is why she came out okay on the other side, because this woman was willing to work with her, for a while we were taking her twice a week. And then we were going to counselling individually, without Maria and she'd stay with my mom while we'd go to counselling. So that was a pretty tough time."

"You were going to counselling regarding Maria's … ?"

"Right, and the counsellor would talk to us about you know the things to do and things not to do and those kinds of things because it is kind of tricky because when she was home she'd talk to us; it wouldn't have been a problem. Now, if you'd come over she wouldn't talk to you. So this was probably the time from kindergarten all the way through fifth grade, basically. We didn't actually get her into counselling until she was seven, by the time we finally found a counsellor. But it was apparent there was a problem and then the school, the school wanted to classify her as a speech-disorder and so, eh, what I did is I did a tape recording of her at home and I took it into one of these counselling sessions that we had at the school and said okay this is her talking, reciting a poem, so I was willing to do that because you know they had their minds made up that it was a speech disorder and we knew it wasn't a speech disorder. So, anyway that was kind of a rough period, she had close friends that she would talk to and, eh, you know in class apparently what she would do is she would tell them what she wanted to convey to the teacher and they'd tell the teacher. So, eh, that was kind of a tough time. But every school year, the beginning of the school year this counsellor would go with us to the school and make an appointment with her teacher at the time and the counsellor would sit down and we all discussed the situation. And finally in fifth grade she just had a breakthrough where they had to write an autobiography and everybody got up and shared the autobiography in the class and the teacher she had that year was kind of a laid back guy." Fred pauses at this point and tears well.

"You know, an older guy, pretty laid back. He had pets in the classroom and all those kinds of things and, ah, it was just, he didn't make a big deal out of it, it wasn't like she's special I'm gonna treat her with kid gloves or anything. He said, 'okay Maria, its your turn,' and she got up and read her speech. So, that was a breakthrough moment for everybody so after that things were more or less normal."

More tears. Fred's emotional connection to this moment is palpable.

"How did you feel when you heard about that, that she'd done that?" I ask.

"Ah, that was just like the most exciting thing that I'd ever heard in my life, it was really exciting because you know it was a long road to that point and ah it was hard for her and we'd be out in public and people would call her shy and that didn't help any, eh, you know, eh, she would need … you know in a restaurant she wouldn't order for herself for a long, long time and she finally got to the point where the waitress would say 'What do you want?' and she finally gave her own order. But that took years."

Father Bonding

Despite his mother and friends from church, Fred's community of support through Maria's formative years was somewhat limited. It is often lonely for parents fighting for a child with special needs.

"Em, so Irene was nursing and she'd be nursing shifts so she'd be gone at the weekend and also in the evenings so you'd be around, you'd be helping Maria with homework." I wanted to know more about the relations between Fred's emotional changes and his connection to his daughter.

"Yes, she pretty much, she was … there was a period in there of a couple of years when [Irene] was able to work an eight hour shift but for most of the time she worked a twelve hour shift so, ah, mostly days so she'd leave early in the morning, sometimes Maria would be awake sometimes she wouldn't and, eh, then she wouldn't get home until eight o'clock at night so sometimes when Maria was younger she'd be asleep by then. She wouldn't even get to see her at all. So I was the only parent that she saw some of those days."

I hesitate. "Would you say Maria's … em, did she bond to you in a different way than she bonded to Irene, given that dynamic? Maybe that is an unfair question."

"No, it is not an unfair question. I just need to probably think about it for a minute, em … Well … I guess we spent a lot of time just playing together and she developed this thing, this game that she called the bed game." Fred chuckles at this memory, "so it was kind of a tag thing and whoever got to the bed and touched the bed was safe otherwise you were free game to chase all around the house and that is something that she came up with on her own so we played that together a lot and you know in the evenings and she'd say, 'Dad, let's play the bed game.' And I remember she was probably eight and we were still playing that and then she got to the point that she out grew that, but …"

"You didn't, though?" Fred and I are laughing now. He stretches his body, easing into an answer.

"Well, it became … after a while I figured it out, eh, so and we spent a lot of time, we'd go to the zoo together, we would go to parks together a lot without Mom. And we'd do outdoor kind of things 'cause that's my thing, so we'd do outdoor kinds of things and there were times when I would take her and a friend usually, em, camping. Irene would be working all weekend so I'd take her and you know go hiking and fishing. She never got into fishing very much, she tried a few times, so those kinds of things."

Belonging and the Perfect House

"Now, you mentioned, Fred, earlier on that the only house that Maria has ever known is the one you live in. How important is that house and that environment for you as a parent? Does that make sense?"

"Eh, yeah, it does but I got to think about it a bit. Eh, well we, we've changed the house around a couple of times. The first time we added on a master bedroom so she could have her own bathroom so she changed rooms at about seven. Em, …"

"It was originally a two bedroom house?"

"Yeah, for all practical purposes. It had sort of additions on the back but you know they were really … one was set up as a bedroom but it wasn't really finished for a bedroom. Not a legal bedroom. Maria was in one of the two bedrooms, the original bedrooms at the front of the house. So she was in there and we were in the back bedroom, just a short hallway away."

"And then you did the master bedroom on the other side of your garage?"

"On the other side of the kitchen. So, then at seven all of a sudden we were across the house from her but that worked out okay. She didn't … you know, the first few days, em, she was a little bit you know, it was different but you know she adapted pretty rapidly.

"So I know at times over the years I've always been the one whose been interested in looking at houses, and I had never planned to live in that house my entire life when I bought it, eh, I just didn't bother checking with my wife first when we bought the house, I just assumed she would do what I wanted her to do." Fred's face collapses into a series of chuckles.

"And she didn't want to move, and we weren't really in a financial position to afford it realistically but, you know, I always wanted to move out to a different neighbourhood, to a quieter neighbourhood, a better neighbourhood. And so we were always looking at houses at the weekend you know, open houses and new developments and I also have an interest in architecture so some of it was that and some of it was I was scoping to find the perfect house to move into and so she kind of grew up with that and there was a point in her earlier life where she enjoyed that. And I remember one time when she was probably 12 or 13 we found a house in the neighbourhood, it was an open house, and we were driving and we all went and we looked at the house and, eh, I told her, 'You know I really like this house, I like a lot of things about it,' and she says, 'Dad, I don't wanna move.' In the back seat, 'I don't wanna move.'"

Fred puts on a pleading voice to emphasize this last point.

"And she … I could see this look of real concern on her face, you know, so the house for her was always a comfort zone you know it was a safe place to go to after a gruelling day at school of being in fear of whatever she was in fear of and then coming home it was a safe place to be."

"So what did that do for you, then, with your aspirations to move to a different neighbourhood, when you heard her say that?"

"Well I think for me that was kind of the eye opener you know 'cause I'd come to terms with the fact at that point that Irene didn't want to move, period. I still was holding on a little bit of hope that I could convince her with the right house but I think by then I was more realistic in my thinking and I realized that financially we couldn't really afford to move to the kind of neighbourhood that I wanted to move to and you know I'd given up on winning the lotto by then." Fred's face folds into chuckles again. He was letting go of his dream-house in a utopian neighbourhood.

"… before it was always oh I am going to win the lotto and then we'll buy a house in La Jolla, you know? I really believed that for a while. So, em, but I don't know. For me the house is comfortable, em, for the last while we've dealt with construction in the vicinity, we've been going through that for going on two years now, but, well, even more than that. There's been on-going construction behind us for a long time, various different projects. So, there, you know, I am really tired of hearing the back-up noise every time a vehicle backs up you know beep beep beep. But I know we are kind of down to the wire now. October is when they are supposed to be finished. But then there is going to be more construction going on with the extension of [Interstate] 52, which is a little bit further away but still you know, it is not that much further away so I know there will probably be noise and dust for another couple of years. So, that I don't particularly like because it interrupts, it interferes with my serenity and being able to go out in the backyard anytime other than a Sunday morning and have peace and quiet and not have a lot of distractions.

"But I am comfortable here, you know, I don't know what it is going to take for Maria to move out, you know, and she has, eh, said that she doesn't want to move out until she has a room-mate, you know, em, she doesn't want to move out so she is comfortable here. And right now she is looking for work. She spends most days at home or she has a job interview or something and she is comfortable there and so I am comfortable there, Irene's comfortable there and it is just we've pretty much got the house set up the way we like it now."

"Were there any significant changes in the house around Maria's growing up apart from when she was seven and the master bedroom?"

"No, no, we did the kitchen three years ago so she was already in college by then so, em, you know she participated and we got her involved in some of the decision-making and things. So, no, other than she always had a say in like paint colours in her room and things like that. I remember times she'd say, well, we had this dirty old beat-up couch in our family room that neighbours had given to us and she said, 'I really don't like that couch anymore I wanna get a new couch,' and she just kept hounding us so about it so, 'its too dark in this room,' so we painted the walls white and then we did get some new furniture so it has always been important to her you know to be comfortable in her environment you know, I don't remember that when I was a kid I, you know, because we moved a lot, we always, if we just move over here things will be better, that kind of philosophy, was my dad's philosophy."

Moving and Staying

"Oh, interesting!" Back to the point of Fred working to not be like his father.

"Yeah, so we were constantly on the move so that is why I thought that, you know, that is what we would do, that is what normal couples do, and, ah, you know she had lived, eh, from her very early years, Irene had, they moved around quite a bit. Her dad was basically a farm labourer who followed the work and they were in the Central Valley for a while and they were back in Blythe and that kind of thing and then he got a stable job, em, she was, before she ever started school so they lived on the same piece of property from that point on, the same piece of property her mom still lives on …"

"Up in Blythe?"

"Yeah, they built when [Irene] was in High School, they built a, eh, newer house because before that they just had a very small, very, very tiny little place which is still there."

"So you are talking about the house and the comfort level of Maria in the house and yourself as well. What about the neighbourhood immediately around the house in terms of raising Maria, how did that work out?"

"Well on the one side, em, there's a family that still lives there now that, ah, had kids, I think three kids, one's five years older than her and then they have a son that is two years younger than her and a daughter that is three years younger than her so when they were growing up they were, you know, inseparable. They were like brothers and sisters practically. They spent a lot of time at her house. She started going over there after school when she didn't want to go to the afterschool …

"For the most part it's a comfortable neighbourhood. She felt safe, being able to play out in the yard with other kids and it was close to the school. It was within walking distance to the school. So she's always felt comfortable in the neighbourhood."

Fred's home is within walking distance of all the schools that Maria went to prior to her undergraduate work at San Diego State University. Despite his wanderlust, and the frequent sojourns to look at other houses and other neighbourhoods, Fred has remained in one place for his whole married life and Maria, who still lives at home, resides comfortably within this familial domain. To ask him now, Fred would say he is part of the perfect domicile, a utopian imaginary of his own creation.

Utopian Imaginaries

> The channel [to the Island of Utopia] is known only to the natives, so that if any stranger should enter into the bay, without one of their pilots, he would run great danger of shipwreck; for even they themselves could not pass it safe, if some marks that are on the coast did not direct their way; and if these should be but a little shifted, any fleet that might come against them, how great so ever it were, would be certainly lost (More, 1715/1885, 23).

Maria Kaika (2004, 283) challenges the reactionary politics that surround the construction of homes and suburban communities in the peripheral Anglo-American world. She sees this as a larger part of a movement – noted in Chapter 4 – towards the domestication of nature and, as part of this, the home becomes "an autonomous individual utopia" (a homestead) through the exclusion of undesired social and natural elements. Kaika seeks to expose "the dysfunctionality of the private spaces where blind individualism can be practiced in isolation." She calls for an end to the selfishness and alienation fostered by single-family homes.

It seems that the reality of Fred's home space as just described is to a large extent one of single-family isolation, but his imaginary begs a closer look at what utopianisms comprise. The ways Fred has come to terms with emotional practices,

I contend, colonize utopian imaginaries in good ways. What I think is lost by the generalizations of critics like Kaika, are the variegated responses to utopian ideals. Fred creates for himself a utopian vision of domestic bliss from which he extends fatherly care to his daughter.[5]

David Harvey (2000) notes that utopian visions of the world are not necessarily bad, but when they fail in practice they leave trails of betrayal, disappointment, frustration and resentment.[6] Paul Knox (1991, 187–8) points out that the US planning profession lost a good deal of its moral authority because of its inability to deliver utopian communities and, as a result, it became more aligned with the development and real estate industries with their rendering of utopian images. They sold a dream and then failed to deliver. This alliance led to a commodification of planning functions, which, in turn, facilitated the privatization and decentralized governance of urban life described by McKenzie (1994) and McCann (1995). What proliferates is a bourgeois utopian desire to secure isolated and protected comforts through propertied individualism (Harvey 2000, 139).[7]

---

5 In *Family Fantasies and Community Space* (1998), I describe the penchant for some young families to aspire to Kaika's utopian domestic spaces that 'set the nuclear family apart'. Exclusive residential communities replete with grass lawns and white picket-fenced houses – but also gates, walls, private security forces and speed bumps – are part of a mythic ideal that is beyond the means of most Californian families.

6 Harvey points out that no matter how well intended, visualized utopias – like More's original conception – are tightly circumscribed spatially and morally. The internal spatial ordering of More's island strictly regulated a stabilized and unchanging set of social process through, amongst other things, a series of exclusions (Harvey 2000, 160). Amongst these exclusions, for example, was a prohibition on marriage before women were 18 and men 22: "… if any of them run into forbidden embraces before marriage they are severely punished, and the privilege of marriage is denied them, unless they can obtain a special warrant from the Prince" (More 1715/1885, 232). It seems then that the moral law of the father (in the guise of the Prince) holds sway in More's island utopia.

7 The appeal of utopia is its promise. A number of the young families in the *Family Fantasies* project told us that they really liked the idea of living in a gated community. When we talked to them prior to the birth of their child, they lauded the notion of a real community where people interacted with each other and talked on a daily basis. Some bought into the idea of neo-traditional residential developments and when we talked with them after their child was a few years old and they had moved into their 'dream' community, a very different reality emerged (Aitken 1998, 129–30):

"Scripps Ranch has everything imaginable – except a theme park – for kids," said one mother. "I cannot think of anything that could improve the neighbourhood." She considers for a moment. "Okay, make it a completely gated community."

"I don't feel comfortable with cars driving on the street at all," said another. "In fact, we're having a gated community. I'm pretty sure it's going in. If that happens, I'll be much, much happier."

I sat with two parents who had expressed interest in a neo-traditional community where people interacted more with each other. When their daughter was born, they moved to this

Utopian promises suggest prefect order and harmony and, in so doing, they anaesthetize us to the messiness of social relations. What I want to suggest here is that another narrative, an affective counter-topography if you will, arises from considering the emotional work of fathering (and mothering). Affective counter-mappings are soulful and their achievement is hard work; they tug at my heart to the extent that I am mobilized to action. Fred, for example, breaks through the placeless and timeless contexts of caring for a child with special needs to create an appropriate place for his daughter and wife. A further example helps illustrate contexts from which emotive fathering arises.

Searching Beyond the Suburban Idyll

Stan grew up in the post-WWII utopian family dream. Sitting on the deck of my home, he begins our interview from this perspective.

"Well I'd probably start with my conception of fatherhood before I became a father or really more the fact that I always knew I wanted to be a father. I grew up in a fairly typical middle-class American family post-WWII and the mom and the dad and the car ... I don't think we ever had two cars but that was supposed to be part of the dream, but we always usually had one. It was certainly a typical family with four of us kids and, eh, we kind of grew up just feeling like this is the way it is supposed to be."[8]

Stan was raised in the San Francisco Bay area. At 22, he got married. He soon realized the inappropriateness of that decision. Children, he felt, would straighten things out.

---

kind of community. I asked them if they ever shared childcare with their neighbours who had a son of the same age. "Oh never, we hardly know them," was their response.

It seems that some parents in the *Family Fantasies* project bought into the idea of community, but what they settled for was security and speed-bumps. The problem with utopian ideals is that they rarely accommodate the complexity of lived social arrangements. This, in part, was the reason for the demise of Robert Owen's New Harmony (see Chapter 3).

8 This is precisely the utopian imagery and timeline that Stefanie Coontz (1992) describes in her acclaimed *The Way We Never Were*. She shows that the ideal of breadwinner father, full-time homemaker mother and dependent children was mythic construction that came after the 1950s. Families of the 1950s were, in actuality, rife with conflict and repression; they were frequently poor and much less idyllic than many assume. Further, Coontz contends, the nuclear family was elevated as the apex of family values only in the late 19th century, thereby weakening people's community ties and sense of civic obligation. Coontz disputes the idea that children are raised appropriately only in nuclear families. Viewing modern domestic problems as symptoms of a much larger socioeconomic crisis, she suggests that no single type of household ever protected Americans from social disruption or poverty. An important contribution to the current debate on family values, *The Way We Never Were* describes a utopian imaginary within which a certain form of family-values fitfully reposed.

Add Children and Stir

"Em, I got married prematurely. I realize that now for sure but I actually realized it soon after we got married. And I kind of had the feeling becoming a father was going to straighten some things out. Make things right."

As part of the utopian vision, children are, for some, a panacea. This was not the case for Stan.

"Yeah. And em, like I say, I kind of felt like that was going to make some things right that I felt were not right in our marriage. And, of course, that didn't happen. It just changed things. It complicated things because now there was another human being that we had to take care of and be concerned about. But as far as my feelings about being a father, I was overjoyed. I remember when Jerry was born the feeling of … it was almost euphoric. I was there after the birth. We had gone through La Maz training so I was very involved in the whole birth scenario."

"You were present for the birth 'cause you did the La Maz?" I ask.

"I was her La Maz coach, as it were." Stan elaborates.

"I can relate." We laugh.

"Yeah," Stan continues to laugh. "And while it wasn't, you know … She had thought that she wanted to do the complete natural childbirth, em, as it got closer it became obvious to her that that wasn't really what she wanted to do but the preparation classes were still good because they helped with the labour and all that. So that is kind of how I felt about the first one. And the first one, while he wasn't specifically planned, per se, em." Another chuckle. "We knew that we wanted to have children and it was just a matter of, 'when she did get pregnant.' We weren't trying to avoid pregnancy. Em, our second child was the only one that was actually planned. We decided that we wanted to have another child and we wanted her to get pregnant so … I remember that whole experience being so, so much different for me in that I was looking forward from the time we decided to have another child until she was actually born. And she was born in the afternoon and there was a song out at the time called 'Afternoon Delight.'"

"Yep, I remember that one."

"Yeah, the *Starlight Vocal Band*," Stan laughs. "I never heard any other song by them since. So she became my 'afternoon delight.'"

"Yeah, and what age were you?"

"I was probably 26. So, and that one was special also because it was a daughter and – we didn't know ahead of time what our children were gonna be, back then they only did that if they suspected there was going to be some problem with the pregnancy, when they do an amniocentesis. So it was still in that era of: surprise it is a girl. So that was pretty neat. Let me think, the third one … by the time we had the third one we had actually had a separation and, em, it was not something that I wanted … it went completely against my grain, my upbringing, my feeling about fatherhood and, em, and just being a husband. The idea of getting married with the thought of well 'am I getting divorced someday' was alien to me. So, em, when she decided that she wanted to try a separation I went along with it but I wasn't happy about it at all."

"And she was pregnant when she asked for the separation?" Stan's utopia begins to unravel.

"Well, we did not know she was pregnant at the time we separated. But, like I say, because I was not into this idea of being separated I was still going over and seeing her multiple times a week and, em, going to see my two kids. And I was just generally miserable at the time."

"So the kids were with her and you moved out?"

"Yeah. It was during that time that she … You know there is a lot of background on my ex-wife that would explain some of this stuff but basically she was in denial about being pregnant. She thought something else was going on and she was just gaining weight and so she was trying to lose weight by eating less and less and then finally I said, you know … and she said, 'well it seems to be centred right here,' and I said, 'I think you're pregnant.' 'Oh, no I can't be,' … 'oh yes you can be.'" Stan laughs again.

"So she went to the doctor and confirmed that she was. And that was kind of the impetus for us getting back together. And it seemed like to me that things were getting better because she and I seemed to be figuring things out and getting a little closer. And now we had three children and, you know, it was exactly what I wanted.

"And it turned out that that wasn't exactly what was going on with her and we lasted about 18 months after Laura was born and we split up, finally."

"Tell me, Stan, about the … you mentioned, em, when Jerry came along that there were things in the marriage that you thought a child could solve. Can you tell me a bit about what those were and then what happened when he came along that suggested to you that there was something else going on?" I wanted to know a bit more about the dissolution of Stan's utopian world-view.

"Obviously I was pretty young and naive at the time. And em, I … Our marriage had already shown signs … we got married too quickly after we met that I started having suspicions early on that maybe she and I were not really that compatible. And, em, so after two years of marriage it was becoming more obvious to me it was going to take a lot of work, which I was willing to do because again, like I said, I was brought up with the idea that when you got married that was that and you did whatever you do to make it work. So I, I guess, the thinking that I had at the time that having this child in our lives made things right or made things better was that it would renew her commitment as well and make her feel and see that, you know, we had a family here not just two people that happened to be married and that was worth working for."

"So Jerry came along and then Hillary came along and you stuck together for how long?"

"Let's see … we were … we stayed together for another two years."

"Where were you living?"

"In an apartment complex over at … No actually we were living in a duplex pretty close to where I live now."

"And how were the living arrangements? Did the kids have separate rooms? Tell me a little bit about the house itself?"

"It was a two bedroom duplex …"

"So the kids were in one room and you in the other except most of the time…"

"Except that most of the time whichever child was the youngest was in our room … for a longer time than I would have liked."

"Right … did you have a crib set up in the room?"

"Yeah."

"Did that create tension between you and your wife?"

"I think it did yeah," Stan chuckles. "Part of that was the constant interruption of sleep. I was working every day and she was a stay-at-home-mom. So my sleep was pretty important to me." More chuckles.

"Yeah. So, em, Hillary was born, then, at what time did the separation occur?"

"She was probably about two."

"And Jerry would be about four?"

"Yeah."

"So where did you move?"

"I got an apartment over here off of El Cajon Boulevard." Stan gesticulates out beyond my balcony in the general direction of the Boulevard.

"So you were pretty close. What kind of … was this like an amicable separation without lawyers involved?"

"Yes."

"So what kind of arrangements did you come to with your wife to see the kids? How did that work out?"

"Well, it was pretty open. I think as she had been the one who wanted to do this separation, she termed it a trial separation which was a term that I had difficulty with … She was really trying to see whether it worked for her living on her own … it wasn't necessarily that she saw it as a precursor to divorce."

"She was living on her own with the kids. You were still the breadwinner and were still paying the bills?"

"Yep."

"So now you've got two apartment rents."

"Yes."

"So how often did you see the kids during that time?"

"Oh, a few times a week."

"Did they come over to your place or did you go over to their house?"

"They came over to my place very rarely, it was mostly me going over there."

"And did your wife take care getting them to preschool? Did they go to preschool?"

"No, not at that time."

"So your wife was hanging around the house during the day."

"Yes."

"And would you see them at the weekend? Take them for weekends?"

"It was never really that I took them separately. We always … it was always me going back over there and being with the whole family." Utopia reformulated.

"So you were doing things as a family with your wife also then?"

"Yeah."

"Okay. And so then Laura came along?"

"Right."

"And right after the birth you moved back in."

"Correct."

"How did that work out?"

"Well, em …" Stan pauses. I use the gap to get some clarification about the ways he relates to his three children.

"First of all, before you get into that … let me ask you this: How was that in terms of … were you part of Hillary and Laura's birth the same way you were with Jerry? La Maz class and all that?"

"Yes."

"So tell me about the emotional connection. You had that with Jerry, you said, was that true also with Laura and Hillary? The same kind of excitement?"

"There was the emotional connection but the difference was that Jerry was the first and he was a boy so there was that feeling of 'I have a son.' With the girls, you know, I loved the idea of the fact that I had a daughter now and I had a second daughter. My feeling about them was different, more like … I need to be more protective of them."

"Mmmh …" I am wondering about learnt and performed gender distinctions in the way children are raised, "and how did that protection, or need for protectiveness, show up? What did you do?"

"Well I don't know if it necessarily affected my immediate behaviour towards them. It was really more, and the reason I say that that was my … my conception as a father of a daughter … as my daughter goes through her life and grows up I am going to need to be protective."

"And you feel that has been true? That that idea has become a reality or have things changed?"

"Things have changed."

"In what ways?" Here comes one possible crux of the disillusion of Stan's earlier utopian dream. And it is about performing gender.

"Well for one thing society changed a lot over the 25–30 year period. At the time that my children were being born the women's movement was in its infancy and a lot of the attitudes were from growing up in a society where women were – from their point of view – maybe a second class citizen, but from a men's society point of view they were protected: we looked after them, we made sure that they were protected. As the women's movement progressed through the period of time when my kids were growing up, em, there was just a lot of that whole idea, that I agreed with, and my ideas and feelings changed about the role of men and women so that as they grew up, and especially as they got into high school and graduated from high school, em, I saw them and I see them as every bit as capable of taking care of themselves as my son is. Em, that doesn't mean that I've completely forgotten the ideas and feeling I had when I was growing up but, em, I don't feel as protective over them in any way more than I do my son."

Time to probe a little and tease out some larger societal issues.

"And how did your role as a father change as, in this 25-year period, as society has changed? How do you see your role as a father changing?"

Single Father

"Well, let me say this first and maybe my situation might be a little unique and, em, four years after my wife and I got divorced I got full physical custody of them. So my role as a father was also somewhat my role as a mother so I have maybe a hard time answering that question as to how my fathering changed."

I recognize this as a crucial twist.

"So tell me the series of events that resulted in you getting custody of the kids." I order.

"Em, she got involved with some guy who got her pregnant and she decided she was gonna keep the child and when the child was born by this time Anne had moved out to Ocean Beach and, eh, it was a crazy time in her life and at that time I had been divorced long enough that I had kind of gotten used to a certain life style or scheme of things and, eh, I was handling it okay.

"So, her mother actually, I mean it was at a point with Anne where she was having so much difficulty with having these four kids by herself 'cause the guy who got her pregnant was pretty much gone that the grandmother, Anne's mother, was picking the kids up multiple times a week because Anne would call her and say, 'I can't stand this, you gotta come get the kids.'"

"And she'd pick up the new kid also?"

"Three kids and a baby."

"So [Anne's] mother was looking after the best interests of her grandchildren to the extent that she wanted her grandchildren removed from her daughter."

"Yep … yep … So I had been talking to friends of mine about it at the time. Friends that I'd known since high school and even before. And they said, 'Well, sounds like that's the way you need to do it' … from my lawyer you know he told me basically that if I wanted custody of these kids I'd have to go into court and prove her an unfit mother and essentially destroy her reputation and character."

"And was that something you were willing to do?"

"I was not. So he had told me if things are with her the way you say they are give it time it will probably happen anyway. And it did." Stan's voice gets quiet and his generally happy demeanour takes a more sombre aspect.

"So it ended up where, em, a very tearful time and Anne and I spent a lot of hours talking about it and she finally agreed to it."

"Without the legal system getting involved?

"Well it certainly was done legally but it wasn't two lawyers fighting it out. It wasn't that she was fighting the loss of custody and hired a lawyer to help her do that ... It was amicable."

"So you ended up with your three kids. Where did you end up with them? Were you living on La Mesa Boulevard at that time?"

"I was at the time, but I realized that I wasn't going to be able to stay living there with three children so I quickly went out and found an apartment in a complex that was very family friendly. Actually, at that time it was still legal in California for landlords to exclude children from apartments so you could find, you know, you'd look through the newspaper you would see: '1 bedroom apartment, adults, no pets'. That would be the phraseology. So I found this place that was family friendly and moved into there and got a three bedroom apartment ... very close to where I live now."

"Now were the kids going to school in Ocean Beach?"

"Right."

"So you were getting custody but they were all moving school to be here at the Spring Valley school district."

"Right, but it was also at the beginning of the school year so I wasn't taking them out mid-year and changing schools ... actually ... yeah, that's right."

"What is your perception, looking back, at how the kids felt about moving in with you and moving away from Mom?"

"Well I've had that discussion with my kids. And, em, especially Jerry realized that Mom was out of control and, eh, of course Laura was still only five years old so she has never really expressed a whole lot of feeling about what that felt like to her, em, but Jerry realized that it was better for them and Hillary has also said that. Laura being only five she came to move with me, to live with me, essentially she grew up with me and while she remembers some things about living with her mother they are just isolated memories, no continuity."

"So tell me about that first Fall – 1985 – where you'd just set up an apartment and you have the kids now and you have them full time: what was your life like?"

Stan laughs.

"I was as crazy as my ex-wife was. It was very difficult. I had a good job at the time and even through the divorce I had a friend who had told me, 'no matter what you do don't let go of that job, you don't need being unemployed and looking for a job on top of everything else.' So that job became kind of my rock. So, em ... you know having that gave me a community, if you will, of resources."

"What kind of community?"

"Well, I had people who I knew at work that I could talk to about what was going on, what my situation was ... They were offering suggestions and offered to help with one thing and another. Plus I was involved with the church – St. Martin's – and actually I'd started taking the kids to church with me on Sunday prior to them coming to live with me and there was a parenting group there that I became involved with so I knew people at the church as well. And that helped me take care of things like afterschool care, things like that."

"So you went through St. Martin's for afterschool care?"

"No, I actually had a friend who worked only part time and was home in the afternoons. So she started taking care of the kids."

"I see. So she picked them up from school and looked after them until you got home …

"… about 5:30."

"And did you then take the kids, cook them supper and look after them?"

"Yeah."

"Were you responsible for getting them all to school in the morning?"

"Yep … And actually when I look back on that period of time you know other than the fact that there wasn't a mother in the house we actually ran a pretty normal household. I know that I did a good job of teaching them the things they needed to know about how to take care of themselves … they tell me that."

Here I get a wonderful picture of a 'pretty normal family' under the tutelage of a working single-father.

"Well, you know, I kind of emulated the kind of things that my parents had done on how to teach kids how to take care of themselves, em, oh boy … I haven't thought about this in a while but like for example I had the kids rotating through the kitchen duties … one of them would wash, one would dry, one would set and clear the table and then they would rotate that job, I don't remember if it was daily or weekly or whatever but I had them doing that, and as they got a little older I had this little bulletin board that was up on the wall in the kitchen – a little cork board – and I had one of these pages out of a photo-album, plastic pages with little envelopes or pockets and you put the pictures in it, and instead I had chore cards and each pocket was each kid and then I'd rotate the chore cards. So each day one of them was responsible for vacuuming, one of them was responsible for making sure the bathrooms were clean, you know, that kind of thing. And I taught them how to do those things and they did them on a daily basis although it was funny because they would, as they got older and they were able to be home before I got home. As I was pulling up I'd hear the vacuum cleaner going on, not off: 'Oh God, Dad's home!'" Stan laughs. "And they'd run through the house getting the things done but, eh, they learned how to do those things and then one of the other things we did was on Saturday mornings we'd bring all the laundry out and we'd dump it all on the living room floor and I taught them how to separate 'em out into different loads by color and fabric and whatever and then each one of them was responsible for doing a load of laundry and I did the rest. So, that kind of thing."

Father as Superhero

A large part of me believes in an alternative to the familial utopian imaginary that Stan brought into his marriage at 22 and the ways he tried to fix that with children. It is an affective counter-topography that focuses on emotions rather than spatial frames. I asked Stan about that.

"So tell me about your emotional connection to your kids at the time. You talked about it being an emotionally raw time for you and you mentioned the connection with how busy you were. What about the connection with your children at that time? How was that changing and evolving?"

"It was over a long period of time, the evolution of being a parent and having your children grow up. Emotionally connected to the kids? At the time that I first got custody of them I had a strong feeling of, you know, it was us up against the world, so to speak. I felt very protective of them. And they became a big part of my identity. I recognize now that I became 'Father Man' …"

Stan and I laugh at the image of father as superhero. He continues, "… and my whole life was around the fact than I was a single father and there was me and those kids. So, a lot of the emotional connection I had with them was just in fulfilling my own needs, which was not something that I ever recognized or thought about back then: that they were doing that … and I am not sure I would have had that strong a feeling about it if I'd been fathering in a nuclear family."

Stan stops for a moment, and then tentatively continues.

"But then things were unravelling for me and again, em, like you just said it wasn't really about being a parent but it was affecting my parenting."

"In what ways?"

"Well … one of the things that I realize now … was the boredom of being a father."

"The routine?"

"Yeah, yeah … It was everyday the same you know. Wake up, get the kids ready to go to school, get everybody out the door including myself, go to work, come home, I made dinner every night. They helped me sometimes but that wasn't really part of the scheme most of the time. And then I would get my bottle of wine and I would go back to my room … I wouldn't necessarily close the door and shut myself off but I mean I was sitting there drinking wine most of the evening and that pretty well provided me some barrier. Eh, so what was happening then I guess was that I was starting to separate from them.

"Jerry at the time was a junior in high school and he was getting ready to take the SAT exams … so he had apparently brought the SAT information to me and I looked at the date that we needed an answer, or fill it out, or whatever … it needed to be back. And I thought okay I can deal with this later so I told him that we'll talk about this some other time okay. Well then he came to me one evening and I was sitting in my room at my desk trying to read a book while drinking wine and really just kind of enjoying the buzz of being drunk and he came in and said: 'Dad, remember you and I were going to talk about the SAT exams?' And I looked up at him and had absolutely no recollection of us having talked about it before at all. And I just looked up at him and he gave me such a look … I just … I was horrified. I felt like I was this tall …" Stan gestures with his thumb and forefinger.

"So, yeah, things were falling apart a little. I mean, actually, when I look back at it I was holding them together."

"Right," I reply. "You were doing a pretty good job as far as I can tell. Pretty amazing."

Broaching the Leaks

Iain Chambers' (1994a) notion of a leaky habitat is appropriate for many of us, and is exemplified quite well by the stories of Fred and Stan. Although there are circumscribed spatial framings – Fred's search for the ideal house and Stan's 'fix it' solution by having kids – there is no common destination. For Fred, utopia came with acceptance that his home of 22 years was good enough, but it was also delimited by the hard work of becoming an advocate for his daughter against the school system. For Stan the struggle was against the ways single-fatherhood framed his sense of self and how that led to boredom. The utopian dream of eventual arrival, a coming home, suggests Chambers (1994a, 245), "now reveals a gap in which some of us begin to lose ourselves." Diversity and fluidity disturb both the passage and the homecoming. Our representations and utopian imaginaries no longer hold (as if they ever did) things together, and those of us who bind ourselves most tightly to their rigging are the ones who tend to sink with the ship. Tracings do not work, we need to make a map that is a counter-topography.

The journey – the departure and homecoming – is caught within a larger itinerary and between other journeys and destinations, which "pose the perspective of an interminable movement, and with it questions connected to a lack of being placed, to the proposal of perpetual displacement" (Chambers 1994a, 245). We never go home because we were never there in the first place. Fred and Stan created homes as continuous placements that evolved from the displacements of older spatial frames. They, like all the men in this book, live in a worldly prose that is way too vast to be their own, but it is also intimately tied to their 'becoming father'.

They live in a space – a spatial frame – that they transform into a dwelling – a home – that is borrowed. This is a more open ended sense of home in a culture that looks to the perspective of leaky habitats.

I glance over at Fred, who is sitting with a pensive look on his face.

"Well Maria is very lucky to have a dad like you Fred."

His pensive look transposes into a thoughtful smile.

"Well thanks, I appreciate that, it means a lot to me."

"How about that cup of tea?"

Chapter 11
Coming Home

> As a space of belonging and alienation, intimacy and violence, desire and fear, the home is invested with meanings, emotions, experiences and relationships that lie at the heart of human life. Geographies of home are both material and symbolic and are located on thresholds between memory and nostalgia for the past, everyday life in the present, and future dreams and fears (Blunt and Varley 2004, 3).

I don't notice the three red steps or the porch. Registered only peripherally, the steps pass under my feet and are forgotten before I enter Tony's living room. I don't notice the colour of the porch or the small carefully constructed bed of plants to the west of the steps and the porch. By the end of my meeting with Tony on that balmy June afternoon, my mind was filled with those steps and porch.

But all that comes later. Tony's children fill my first few moments in the house. This is a place of children and their growing up, although on this afternoon the house surrounds only Tony and myself. I feel the love and warmth of the place: it is silent today but nonetheless exudes the possibilities of children running and playing, a cacophony of young voices. The house is cared for in a haphazard way: clean but only partially picked up.[1]

Tony raised his three girls in this house. He was raised in this house. One of his daughters recently returned to raise her child in this house. Tony was a child and is now a grandparent in this place. He is deeply involved in the surrounding African-American community, although today he says it is much more mixed ethnically due primarily to an influx of Latin American immigrants.

I am seated in his living room and I can see in past the dining area to a room with broken lathe and plaster; scattered tools suggest some remodelling going on in the back of the house. This is a place of becoming: an evolving 'classic Californian bungalow-style' cottage of maybe 1,000 square feet.

When Tony begins talking about his children I get a glimpse of the energy I sensed as I walked in the door.

"They were fun kids. They had expression. They had attitude," Tony remembers. "They had energy. They were just bounding around. And then I think when [the last one] was born was when we moved into this house for good."

---

1 Tony works at home: he created a company some years ago that brings together workers with particular skills and employers with particular needs. In control of his work environment and focused for the most part on contracts with the eastern seaboard of the US, Tony particularly likes that his employment is done for the day by 3pm.

Figure 11.1 Tony's red porch steps

Tony's head inclines from side to side as he takes in his home. This is an important place for him. I wanted to hear more about what constitutes that importance.

"But you moved back earlier than that," I question, "to be close to this neighbourhood and your parents?" Tony's world is circumscribed by very specific familial and community ideals. He believes in family, in the connection between generations. He is deeply concerned with changes he has seen in the status of African-American fathers. Tony's involvement in a variety of community programs, including a neighbourhood watch that he started, are part of his struggle to help this and other African-American communities.

The last several chapters moved through travelling stories and connected them to variegated notions of fathering that circled around the home place. With this chapter I want to add to that discussion some cautionary notes on exclusivity, racism, imperialism, the subject-self and domesticity, moving to a consideration of home places and community relations.

I take some of my inspiration from a comment by Paul Basu (2001, 335) who, in his essay on identity and the Scottish diaspora argues that "[hunting down home] … is *not to search*, [but] in part, at least, to seek that which is felt to have been lost." There are, perhaps, parallels between the African and Scottish diasporas and how those play out in the colonial and imperial histories and geographies of those two nations.[2] What I offer here, I hope, is a rendering of home place that is connected to the placing of fathers more generally with an important nod to larger contexts of diaspora, domestication and imperialism. Tony's story is focused on the context of place and dwelling and, I think, is connected (in different and complex ways) to John's fierce defence of hearth and home in the introduction. In both contexts, a Heideggerian sense of dwelling dominates a sense of family and self in an open (hospitable rather then closed and exclusive) way.

Mobile Geographies of Home

In previous chapters I do not address the specific spatialities of home, focusing rather on fathers' movements through, away from and back to domesticity. I am concerned with mobile geographies of fathering in this chapter also, because I do not believe that it is possible to stop moving and searching. Nonetheless the specific spatialities of house and home are important and are studied in recent years from a number of interdisciplinary perspectives (cf. Chapman and Hockey 1999; Cieraad 1999; Miller 2001; Dwyer 2002; Hyams 2003; Blunt 2006). In this chapter, I want to continue the last chapter's beginning discussion on the political significance of domesticity with a focus on spatial intimacies and communal households, and the ways that dwelling evokes a sense of belonging and being. As suggested in the last chapter, geographies of home are not tied to the house or confined to a particular location but transcend scales from the domestic to the global, both materially and symbolically. This, of course, elaborates important ties to imperialism and colonialism, and so, with this penultimate chapter I hope to bring together many of the political themes of previous chapters.

Blunt and Varley (2004, 3) note "[t]he everyday practices, material cultures and social relations that shape home on a domestic scale resonate far beyond the household." Globally variegated constructions of ethnicity, race, gender and sexuality are represented through and performed in domestic spaces. Everyday practices of cooking, washing and cleaning within the domestic sphere, and the actions of building, extending and remodelling dwelling spaces have much wider implications. These practices are situated within a range of complex and contradictory experiences, associations and meanings. They construct a politics that is at times open, surprising and hopeful and, at times, enervating and constraining.

---

2 I do not want to make too much of the connections except to note that the Highland clearances that forcibly moved many Scots off their tenant holdings and on to the Americas happened at approximately the same time as Africans were forced into slavery. Each scenario was connected to larger forces of global economic expansion, and the two nations were curiously connected via the infamous triangular trade. Using the already existing textile industry, for example, entrepreneurs in Glasgow profited hugely from imports of cotton from the American colonies prior to 1776 (at a time when the clearances were at their height).

Dwelling as a Political Event in Space

> These [houses] are now good only to be thrown away like old food cans (Adorno 1974, 38–9).

In his writing on belonging, Adorno is making an emotive claim that "the house is past" from a place – situated after WWII's decimation of European cities – where there is an impossibility of 'feeling at home'. From this place, he suggests that house, home and domesticity are a manifestation of larger national and imperial projects. Of late, from this and other sources, a number of theorists set up dwelling and place-belonging as cautionary tales.

Some of this critique comes from a fundamental dismissal of Heidegger's ideas of being-at-home-in-the-world as "implicitly romantic … in dangerous ways" (Thrift 1999, 310), or "sinister (nationalistic)" (Cloke and Jones 2001, 661).[3] Heidegger's later works, such as *Building Dwelling Thinking* (1993/1951) situate 'dwelling' as a key part of his critique of Western metaphysics. Reacted against by Frankfurt theorists like Adorno, Heidegger's work is sometimes seen as an emblem of "regressive nostalgia," and "reactionary nationalisms;" home is often "dismissed as an embarrassing backwater of unemancipated feelings" or, worse, it is used as a "breeding ground of repressive and oppressive politics" (Bammer 1992, x). Harrison (2007, 642) concludes that dwelling for Heidegger is about enclosure, autocracy and self-sufficiency. From ideas such as this, Massey (2005) argues that there is an exclusivity tied to notions of place and belonging because they establish community on the basis of sameness. Heideggerian conceptualizations of place, she argues (2005, 183) are "too rooted, too little open to the externally relational." In a similar but more polemical vein, Bonnie Honig (1994, 570) – seeing home as totalizing, privileged and imperialist – urges resistance against its "seductions" because "the unitary self is founded on the exclusion of difference" (Varley 2007, 3). A politics of home based upon nostalgic appeals to unity and privilege, argue post-Marxist and feminist theorists, is oppressive and debilitating because it suggests a closed system that denies community and difference.

Equally problematic is a nostalgia-streaked sentimental representation of past domesticity that never was, as suggested, for example, by Coontz (1992, footnoted in the last chapter). What is interesting about some of these representations, argues Rosemary Marangoly George (1998b, 6), is the elaboration of economic, racial and gender arrangements that need to be in place on a national and international level "… before respectable homemaking is successfully achieved."

The romanticism and exclusivity of houses and the domestic is exacerbated by the rise of implicitly racist commodity fetishism in the 20th century. In a classic indictment of advertising that lauded clean, white bodies and spotless kitchens,

---

3 Of course, Heidegger's involvement with National Socialism and the link between his philosophy and incipient totalitarianism is an important part of this critique (Harrison 2007, 627).

Ann McClintock (1995) shows how the cult of domesticity became indispensible to the consolidation of Anglo-American imperial desires.[4] Racism and domesticity is more than just about creating white bodies and unblemished kitchens. The question of who does the cleaning and gardening turns today, as it has on the past, on a racialized division of labour. In *Domesticity and Dirt*, Phyllis Palmer (1989) notes that foreign domestic labour (both legal and undocumented) is central to US childcare and home maintenance. In raising this issue to a global context, Cynthia Enloe (1990) and Gerry Pratt (2004) note that there are entire countries in the global south (such as Sri Lanka and the Philippines) where economic stability is largely dependent upon migrant domestic servants – usually women who contract out their labour as domestic maids in Europe and the Middle East. Pratt (2004, 167) calls this a "gendered loop" in which the labour of working-class and racialized or immigrant female labour is domesticated to support the reproductive labour power of employed middle-class women. In citing this phenomenon, George (1998b, 8) suggests with Adorno that there are some forms of domesticity with which we must dispense.

Rachel Bowlby (1995, 77) argues that the "rejection of domesticity has seemed a principle, if not *the* principle, tenet of feminist demands for freedom. The home figures as the place where the woman is confined, and from which she must be emancipated in order for her to gain access to the world outside that is masculine but only contingently so." This is the basis of spatial entrapment theories that were prevalent in geography in the 1990s (cf. England 1996). In this configuration, the excluded other is not necessarily 'out there'; she is confined within a home space. And yet, later in her essay Bowlby suggests that the domestic is also a

---

4 Anne McClintock's (1995) classic discussion focuses on dwelling, advertising and commodity fetishism. The commodity fetish became a central form of industrial enlightenment and sought to hide, through cleanliness, what liberalism would like to forget – that the domestic is political and the political is gendered. McClintock points out that the economic value of women's domestic labour could not be admitted into a male rationalist discourse and so it was disavowed and projected onto the history of the 'primitive' and the geography of empire. McClintock broadens this argument with the point that commodity fetishism is about exclusivity and, more insidiously, it moves racism from the scientific to the market. She uses the example of early Pears soap advertisements (some of the first commercial adverts in the US and UK) that suggest cleanliness is next to godliness and cleanliness is about transforming dirty, black bodies into something that is clean and white. Imperialism gave shape to the development of Victorian domesticity and the historic separation of the private and the public. The point I want to make with McClintock in this chapter is that colonialism took shape around the Anglo-American invention of domesticity and the idea of the home. Through the mediation of globalized commodity spectacle, domestic spaces became sexualized and racialized, while colonial space, both at home and abroad, became domesticated. A mass global marketing of empire as a system of images became inextricably wedded to the invention of domesticity, so that cultural histories and geographies of imperialism cannot be understood without theories of racialized and sexualized power in and through domestic spaces.

place of sanctuary, a safe haven against a menacing world. The irony of this turn is important, as it is the tension that creates an important 'event in space' from which possibilities arise.

From Derrida (2001, 16–17), dwelling and residence are fundamental to the way we relate to others "inasmuch as it is a manner of being there." The contemporary tension that surrounds notions of home and domesticity comes from scholars such as Iris Young (1997), Gerry Pratt (2004) and Anne Varley (2007) who note that along with the exclusivity, sexism, racism and privatized violence that are part of home spaces, dwellings nonetheless provide a valuable material space that supports individual and collective narratives of identity. "Home is the space," argues Varley (2007, 4) "*both* of self *and* of the other; but this ambivalence is denied by too insistent a focus on the exclusionary home." Varley goes on to suggest that other approaches to theorizing the self may allow thinking about home in ways that do not become "trapped in the binary of exclusionary or idealized space" (Varley 2007, 12). One possible encounter returns me to Kristeva's notion of the abject (Chapter 4), which ends with the recognition that the externalized excluded, maligned other is, in actuality, an internal object. Similarly, Donald Winnicott's (1971) notion of transitional spaces recognizes the complex, non-binary connection between internal subjects and external objects. Transitional spaces are spaces of play that bear close resemblance, as a Deleuzian double-articulation, to Lefebvre's spaces of trial (Chapter 9) in that they are about the creation of the subject-self through fluid, recursive processes involving intuition, experimentation and play. Rather than seeking Le Febvrian categories of relations, Winnicott illuminates infinite possibilities by attempting to describe the creative processes through which individuals establish perspectives that reconcile the inner realities of self and the external realities of society, culture and nature (see Aitken and Herman 1997; Aitken 1998). Varley (2007, 12) argues that renderings of this kind provide a richer account of the spaces of subjectivity than the "frozen geometry of inside self and outside other" that inform critiques of home. Ambiguities and tensions are embraced as home becomes both an exclusionary space and a space of mutual recognition, and in its totality the recognition is always a partial and asymmetrical event.

Following Ulf Strohmayer (1998), Paul Harrison (2007, 627) makes the claim that dwellings (and homes) are 'events in space':

> To invoke the concept of dwelling is always to attempt to re-call, to restate or rephrase, an *ur*-concept; it is to describe an originary spacing. An originary and thus potentially immemorial spacing in that the knowing conscious subject will always constitute its distances, perspective, gaze, or narrative *from* the intimacy of dwelling.

From this sense, Young (1997, 159) argues that home carries a material anchor for agency and identity. As an event in space, homes are, at one level, an extension of bodies. They are a repository for memories that anchor a sense of self. They are about domestication in the sense that repeated actions and rituals (meditation, cleaning, birthday celebrations, personal hygiene, watching television) become a slow sedimentation. Dwellings are a sedimentation realized through bodily functions and motions.

Home is a material anchor in a narrative sense also. The home and its objects "sediment personal meaning as retainers of personal narrative" (Young 1997, 150). These stories link the past and present to a hoped-for future, and by so doing further create an event in space. At the very least, seeing home in this way brings the event into thought from a variety of different modes of encounter. Class, race and geographic location place heavy inflections on domesticity and yet, George (1998b, 7) points out that "like love, childhood, and death, the domestic is seen to transcend all specifics or rather to blur distinctions in the warm glow of its splendour." And this splendour rises in the form of critique and transformation of thought.[5]

Seen in this way, and drawing on Emmanuel Levinas (1969), Harrison (2007) argues that dwelling is constituted as open and unfinished, an event in space that is co-constituted through self and other to the extent that binaries lose significance, exceeded by the event. For Harrison (2007, 643), Levinas' work on dwelling is a recasting of "the relationship between inside and outside, interior and exterior, such that it is not based upon a logic of sovereignty, exclusion and absorption but on heteronomy, differentiation, and responsibility." Harrison goes on to conclude that opening up dwelling as an event in space further opens the terms identity, community and subjectivity because it focuses on the issue of the relational, "… or rather, of the spacing of the relation."

For me, and Tony, dwelling works in thought beside ideas of identity, community and subjectivity. It is about the spacing of fathering relations.

The Spacing of Fathering Relations

> We must leave home, as it were, since our homes are sites of racism, sexism, and other damaging social practices. Where we come to locate ourselves in terms of our specific histories and differences must be a place with room for what can be salvaged from the past and what can be made new (Kaplan 1987, 194–5).

> It is only through leaving home that an understanding of the complexity of social location is attained (Pratt 1998, 19).

5 For example, criticisms of home as an oppressive and apolitical space provoked a vitriolic response from black feminists who argued that such positions emerged from privilege that spoke against the safety and security of home spaces for blacks in a racist society. And so, bell hooks (1990) famously describes the sanctuary, love, affirmation and resistance that comprised her grandmother's home in a fiercely racist community in the US South.

If the emotional work of fathering necessarily comprises a search, then surely part of the journey is to seek that which is perceived as lost. Tony travelled extensively as a young man, taking a variety of different jobs. When his first child came along he knew it was time to return to San Diego, to his neighbourhood, to his community and to the family that had raised him and his brother and sister.

Tony tells me about what it was like when he arrived home with his children, and how that move (and his relations with his wife) created a space for him to get involved.

"And so we, we're all in this house and it was a lot different then and they had bunk beds in the bedroom and I think that is when I started … I really started getting involved … And it was kind of strange because [my wife] still had that fear of having kids. Her upbringing was not very healthy in that area, from a child standpoint. Very abusive background."

Child Relations

Tony's brow furrows. He is in his 50s with virtually no grey hair and his body (like his house and his children) is well kept and looked after.

"And so my sense is – and I can look at this now in retrospect – she had a lot of fear going on about being a parent. And what that fear looked like is that she was very hands off with them. So there were probably periods of time in which the kids weren't supervised or managed as well as they should have been. But nevertheless I'd go in and I'd take 'em up into their room and read 'em stories. And they still remember those stories. You know? They were mostly adventure kind of books. You know, the *Chronicles of Narnia*, em, all of that … that was a big book. I'd read only for about 10 minutes – enough to get their whistles wet, so to speak." Tony smiles, "They would really get hooked into it and then I would cut it off just to maintain some enthusiasm for the next night. And that really became a ritual around here.

"And then Adriana came. She is the baby, she is the youngest. She is now, I think, she is 22 years old now. And that, that really complicated things … two children were manageable, but three, well, you know!

"And so we traded-in the car and ended up getting an SUV and started loading it up with everything and going down to the beach and everything was a big production because we had diaper bags with Adriana. We had the toys for the other two you know and all … it was like an entourage, a big production, to go anywhere. You know, we'd go up to *Disneyland* and it was fine with two because I'd have one and [my wife] would have the other and then that third one, which always would end up being Katherine because she was the oldest and she didn't need as much hand-holding as the other two did.

"So the stories I was reading continued, em, you know it was … I generally, I started to get … to get an attitude about them that… My job as a parent, what was that? You know, and I still think about it and I still hold fast to it: The majority of the job as I saw it of being a parent was being a teacher. Teaching them things that they would have to learn to become an adult, a functioning contributing adult, which covers everything:

> it covers compassion;
> it covers discipline;
> it covers love;
> it covers setting goals;
> it covers responsibility.

"It covers not only being responsible for chores and things around the house but also being responsible for your behaviour, you know, and admitting when you are doing something wrong and it is just that whole gamut. And I've seen it cover the whole gamut, and the problem I had was with consistency, making sure I get it all the time and the other was when I was being over-bearing.

"Because at times I was probably pretty – I was – controlling. In a lot of areas, and that was more justified then than it was in the later years with the girls."

Community Relations and the Familiar

Tony's narrative weaves from relations with his children as they grew up to relations with the community, and how he channelled his children through familiar contexts.

"And then they all went to St. Rita's, the Catholic school up here, in the 'hood … from elementary school. And it was close and I went to school up there – so I knew the Monsignor and I knew a lot of the people – and I wanted them to get in a Christian environment at least in their formative years. And I really didn't trust the public schools, em, here in San Diego. And that was primarily because I went to a public school in my later high school years and it just didn't seem as strict: it didn't seem as disciplined and as nurturing at the same time as the Catholic school was. At least, what I remembered. And then when I'd go to the Parent/Teacher meetings and worked Bingo and all that stuff then you got used to the system up there and it hadn't changed much except they didn't have nuns, they had the lay teachers, which is probably a good thing. A couple of the priests that I remember were still up there.

"So, I have the kids, my wife Jasmine would work some strange hours as a registered nurse and so I would pile the kids up and, well, what the routine was – and we laugh about it now – I would make sandwiches for them probably for the whole week, on like Sunday, and I'd put them in the freezer and I'd make the lunches and we'd be eating breakfast and Jasmine would already be gone, I think. I remember that we'd have breakfast and then – early in the morning – and you know they'd put their uniforms on and I used to brush their hair. And now they laugh that I used to make the ponytails so tight that their faces would be, you know pulled back and I'd slop gel on it and we'd be – phew …" Tony motions with one hand sliding over the other, "… out of the house.

"And we'd be going over homework and [talking about] what's going on in the day and make sure they have their lunches and the whole nine yards and I'd drop them off at Blessed Sacrament and then I'd shoot up to Sorento Valley and go to work. And this is pretty early and this is before school would start so they were going to the [early morning] programs before school would start. They would have all the kids in one room and they'd make sure they were okay and then they'd escort them to class from there. And then after work I'd come back, pick 'em up and we'd do the routine. We'd do homework. We'd come here, do homework before any TV got put on. Ideally I'd try to have them do homework because they'd stay in the aftercare daycare as well. Try to make 'em do their homework there. And then I'd go over it with them and we'd have dinner and play around and that was the routine. For what? For years!"

Divorce and Complex Relations

"Em, and again my whole theme was trying to teach them things. You know, about then I think when Adriana was about eight … By this time, by the way, Jasmine and I had started getting paperwork for a divorce … and a big custody battle and the whole dang thing. And so at about 7 o'clock in the morning Jasmine had the other two, I had Adriana, and we're driving to work and em …"

I interrupt. "Are you still living together with Jasmine?"

"No. She was somewhere else. Adriana was with me … you know we had switched off: three days off and two days over here. The other girls were at Mom's house."

"And then you switched and the two oldest came over here?"

"Right, or sometimes … I think by that time, Stu, we'd … it was pretty much up to them. You know, so, they would go through guilt about who to stay with. They didn't want to leave their mom alone because they knew she'd be lonely and they didn't want me to be alone and, eh, and I think … in fact I know the divorce was pretty hard on them because, you know, we had from their perspective, outwardly, a very happy – you know we argued a lot towards the end but we did things pretty closely as a family most of the time. And the divorce was for a whole lot of reasons. One of which was certainly my stuff and then the stuff that she had. I mean it would have been tough to stick together.

But you know, seven o'clock in the morning driving to work, getting ready to drop Adriana off: I rear-ended this car. Luckily nobody was seriously injured but I had a little cut over my head and so I got thrown in jail for a couple of … you know, child endangerment … well for about 12 hours and I came out again. And needless to say [this was] during the custody battle and [it] went to hell and a high basket. So I went through a fair amount of time – a couple of months – more than that, probably about eight months with supervised visitation. It really devastated the kids because my time with them was reduced tremendously. I, of course, did not like it then but in retrospect it was probably … it was the best thing to do because I was able to go through the divorce and we ended up with 60/40 and

Jasmine had the kids 60 per cent of the time I had them 40 per cent and we em, you know, we still had these fights you know and court you know and I look back on it now and it was just such a waste of time. We spent a fair amount of time fighting over who was going to take care of the kids. And they seemed to have gotten through it."

Dwelling and Familial Propinquity

Tony continues with a discussion of life after his divorce. The spatiality of the family, now apart (and yet also connected and evolving), is nonetheless important.

"And we stayed that way for a long time and oddly enough as the kids got older they started to gravitate back over here. Back to this house. And, em, I think Katherine was the first one to actually move in and live with me. She was the oldest one, I think she was about maybe 14 or 15."

"And Jasmine is living close by?" I ask.

Tony picks up on my use of the present tense.

"Em, Jasmine died about a year ago. But, yes, she was living in Lemon Grove. So she was still pretty close. The proximity was always close. I made sure that I was really close to them and Jasmine made sure that she didn't move too far away in general. I think the furthest away they were was in Tierra Santa, which wasn't that far. Em, but Katherine moved in and then the other two eventually moved in with me.

"This, of course, presented its own set of problems … with three girls."

I wanted to get a sense of what kind of family the house contained at this time, and how Tony was supported in the community. What he has, it turns out, is a communal household in the sense described by Gibson-Graham (1996; 2006).

"Now is your mom living here at the time?" I ask.

"My mom was living – is living – about ten minutes away and, em, she would help out a lot. I did not really ask a whole lot of her but she was very … you know, I'd always drop the kids off when I had to go do something if Jasmine did not want me to leave them with her that day or that couple of hours. And I wanted to make sure that they had a healthy consistency with their grandmother and that worked out well, you know, she enjoyed having them around. She is very nurturing. And she would always take care of the kids, treat them as much as possible and take them places and that kind of thing.

"You know? And just raising three girls.

"I mean, even to this day, I've got … two of them live here. The youngest, Adriana, lives with my mom and Buddy, my stepdad. She lives with them because they've both had strokes and she takes care of them. She hangs out and makes sure their needs are met. And then Leena is now married with a little boy of her own, Daniel. And her husband's in Iraq. So she is staying here for about another year. So, Katherine was off doing her thing and then moved in because her living arrangements weren't comfortable for her about two years ago. So she moved in under the auspices of being here temporarily and two years later she is still here. And, you know, they are old enough to appreciate better things now.

"I have backed off considerably the teaching aspect of it, because I figured they've learned from me almost everything they're gonna learn and now it is just basically up to them to do it. They know what to do:

> they know to clean up their rooms,
> they know to do basic things,
> they know
> they know how to do things.

"Now it is just a matter their maturity level allowing them to do that … motivating them to do that."

I look quizzically at Tony's reference to maturity. I do not say anything, but he catches my glance and elaborates. For Tony, like Rex (Chapter 9), there is a connection between teaching and emotions.

"And what I said earlier about them, eh … the divorce having an effect on them I look at that now and they're, … I think they are all a little emotionally behind their age, you know, because I think they went through a period of time that it was pretty rough and they probably stopped. They were in a survival mode at that point rather than growing. You know they were just thirsting for love and security and stability and all of those things and I think they stopped at some point of learning, and I see it now more than … more than ever. But I mean that is sort of an abridged version of what it was like and now its, its fine. We still have our disagreements, even more so I back off and I am more aware of my presence with them and you know, I mean … I don't have a whole lot of rules around here even though they are around here. They don't pay rent. They eat everything out of the refrigerator. But there have never been problems. I think the biggest you know, they had … they went through a period of time… and this is funny, almost to a calendar every six months we'd have a crisis. You know, em, one of them getting into trouble at school you know. I think Leena, the worst thing she ever did … I got a call on Friday night one time because, eh, she was picked up for shop-lifting in Merwyns, down in Bonita Plaza, you know, stupid stuff, and she was under 18 so that wasn't a big deal and, em, and but none of them have ever been problems at all thank goodness."

Community Consolidations: Dwelling, Family and Open Racial Spaces

I wanted to know more about Tony's connection to his community and how that revolved around his house and family.

"When you were talking about your kids going to the same school as you went to – and you've lived in this community now all your life – what is it about this community that is attractive to you? What is it that makes you feel comfortable as a father raising three girls? Or not? Give me a sense of this community and what it is like to live here for you? For you as a father? For your kids?"

"I think one of the things is that from a financial standpoint the house is already paid for, it has been in the family, so I am comfortable around that, would be one of the reasons. The other reasons are that – not primarily by design, but certainly was part of the equation – the kids are bi-racial, their mom was white. And, em, most of the people over here are Latino and African Americans. And I went to a high school that was a Catholic and that was all white. At that time St. Rita's was predominantly white em and I wanted them to sort of have a balance of feeling comfortable dealing in any culture em because a lot of their schools after St. Rita's when they went to Blessed Sacrament I think they were the only black kids up there. And that was pretty rough on them, I mean they were called a lot of names and, em, my kids are very fair skinned and they … they …"

"How did that affect you as a father? Hearing them coming home from school, after being called names?"

"That hurt me, as it hurt me when I was a kid and was called the same names and treated differently."

"Was this more an African American neighbourhood when you were a kid here?"

"Ah, yes, but there were more whites around than there are now. But em, I felt two things: I felt hurt, got angry with the parents because I think this kind of stuff is learnt, it is not, it is not … we are not born with it. You know, and it was hard for me to trust a lot of the parents there after these names and especially when I found out who was saying it. Em, and in Catholic schools this was a pretty small school so everyone pretty much gets to know everybody else and to my knowledge other than what the school had done … what limited influence they had … parents did absolutely nothing about it. And these are Christian, Catholic parents not dealing with their kids even before and then not dealing with it after it was done. And kids can be cruel so I am not just blaming the kids. Probably they had heard from other aspects as well and just brought it in. And then the other thing is I started then telling the kids that, eh, that is the real world. They are going to face that kind of prejudice all throughout their lives and that life isn't fair and that they just have to *keep it moving,* so to speak.

"I think the one that it affected the most was Katherine probably. She was the oldest and she was the most, and is, still the most sensitive out of all three of them. And that had, in my judgment, that had a big impact on her self-esteem. And I think to some degree it still does because she still brings it up sometimes. And Katherine is 25 years old. So, you know, my sense is that that was an emotional scar for her, a big one, and that is just my opinion. You know I talk to her a lot and I talk to her and she brings this up sometimes still. And she remembers the names that they used to call her, you know, 'monkey' and all kinds of other little things and, em, and I think she still has some anger about that, em, she remembers who did it, and once in a while because this is San Diego she runs into them sometimes, and I don't think it is a too comfortable situation for her. She still gets pretty pissed about it. But that is all I can do … just acknowledge the feelings and then try to give her some tools to get through it. Because it is what it is … that is the way it is going to be.

"You know what is really funny is that if you were to look … if you were over at our house on Thanksgiving or at Christmas you would see every frigging race and culture sitting at that table. Because Adriana, my youngest, loves black men. Katherine loves white men. And Leena is married to a Latino. You know, and it is just so funny to sit and watch them and say, woo …

"I am proud of … I am glad that they don't have, you know, any colour biases that I am aware of. So you know that is, I have to take my hat off to them that they stay that way."

"So what other aspects of the community do you think were positive to your kids growing up?" I ask. I am getting a good sense of the household and want to broaden that out beyond the schools to the larger community.

"You know as bad as this may sound I think they've grown up looking at people around here and have learned how they don't want to become. Em, and they haven't been able to get into the reasons why people are hanging out on the street em, why they are not working, but that is what they see. And, em, there is not … there are a couple … we've got neighbours on the right of us and neighbours on the left … immediately to the right and the left who are just excellent. And all, and all this community may look a little rough. And when I say this community, I mean this street. And em, we all pretty much know each other and we look out for each other so they have that sense of community from looking after each other. Because it is sort of like the good guys and the bad guys. And the good guys are all supportive of each other and we sort of have each other's back. We have a neighbourhood watch here so they got a chance to be involved in that when we started that.

"The kids see me talking to different people in the neighbourhood and they know, I mean, they see me as the mayor of the neighbourhood. You know, and ah, and so I think what they've learned you know, I haven't thought about this, but I think what they've learnt is a sense of community because there were periods of time in which we left here and rented this out and then moved back. And so we have lived in suburbia – quote/unquote – and that wasn't anyway near as friendly."

"Where did you move to?"

"Um, up to Spring Valley."

"And what time period was that?"

"Oh man I think … that was when we first moved back here and I think we moved over here when Adriana was a little older, was about eight or nine. Right after the accident, right after the divorce we sold that house and then moved back here. Because we had lived here before, bought that house over there and then when we split up moved back over here."

"Okay, so when you were living there you rented this place out."

"Yes, that's correct."

Dwelling and Father as Events in Space and Time

As our interview draws to a close, I want to come back to the house itself. I want to understand more fully the palpable warmth and security that I feel. I wonder how much that is about memory and generational living.

"So let me ask you about this house and your relationship to this house 'cause it is not usual to have a generational house here and, em, what does this … you talked a little bit about … and your kids have come back here too and I can tell you are quite excited about that. This is something that you are not unhappy about: What is it about the house and about the space here that … you talk about comfort but is there more than that going on here?"

"I think it is a sense of family," replies Tony with no hesitation and then goes on to connect the house with the street and the neighbourhood. "You know, this house has got a lot of history, em, and a lot of good memories, you know? A lot of the kids that I went to Elementary School and to some degree High School and just playing on the weekend, are still in this general area. In fact, we play golf together. I play golf with three guys pretty regularly that I grew up with when I was even younger living close by here."

Tony thinks more about what I like to call the quintessential historical and geographical coordinates of the house. A particular day is steeped in his memory. Tony was about seven or eight; it is the day that his dad died; it is located in the geography of the front porch.

"My mom and then my biological dad … were here only for about a year or two and … And my dad died. Very unexpectedly he was in his sports car, he drove the Tijuana to Ensenada run that they go back and forth with a navigator and everything … And on a Sunday he got into a wreck and died, suddenly.

> I had just seen him that morning.
> And I remember sitting on those stairs out there
> and begging him to let me go with him.
> You know, just sobbing.
> He would drive off and the tears were rolling down.
> My mom would pick me up and bring me in the house from the porch and then
> …
> That was on a Sunday morning
> and on the Sunday night we heard that he was killed.

"And, em, that just had an earth-shattering effect on the family. Because my dad was a good guy. You know, he would come over here whenever he could and we'd go over to his place and he would take us places and he had cool friends, you know, and drove around in this nice Jag, you know, and, eh, would teach us things: working on cars, keeping the cars clean and, em … And that was a really big void, not having a dad. It hit us all pretty hard."

Tony looks out of the window and remembers some of the things he used to do with his brother after the tragic death of their father. He talks about connection to place, to leaving and to returning, to place-based nostalgia, to searching and becoming.

"We'd go in the canyon before these houses were made. There was a canyon over there and I'd take the dog over there and we'd hang in the canyon for a while and catch frogs and fun things, you know, and, and, and then the kids grew up and I grew up and went away to school and came back here and gravitated back here, you know, so a lot of it is just a comfort level. And the neighbourhood was nicer then. It didn't have so much of a drug issue that now permeates a lot of this whole place. And I know people, there was familiarity. And you know I would go and live in a different city in a different state for X amount of years … and this house was always here. Sometimes it was vacant or it was being rented out. But we would always …

> I would always gravitate back to San Diego
> I lived in Louisiana,
> I lived in Denver,
> I lived in San Francisco,
> I lived in Texas.
> And I would always gravitate back here.
>
> I don't know if it is a cultural thing
> but for some reason …
>
> and my brother and sister never moved away,
> they've always been here in San Diego
> you know, and, so you know.

"I, I, I think what I enjoy having when they come back here – the kids come back here … they grew up in this house. You know, as I grew up in this house. You know and I was just sitting with Leena out – as a matter of fact a couple of days ago – and she was sitting on the porch out there with Daniel, my grandson, and I was washing my car and she was just – Daniel is nine months old so he is not going to respond – and she was saying,

> You know Daniel,
> your Grandpa sat here on this red porch
> and he grew up here and I sat up here
> and now you're sitting up here.

"You know, and that is special. You know, and it is rare. And I take that a lot for granted. So, and they always, even when they are not living here, they all got keys to the house. You know? When I least expect it they are pulling in the driveway and coming on in the house. You know? Adriana does that now, she is the only one who does not live here and you know almost every day and I see her truck, she comes up here, snatches the dogs, takes 'em to the beach and comes and drops them off and hangs here for a while. So, it is all good."

Figure 11.2 "He grew up here and I sat up here, and now you are sitting here"

Moving On

In past chapters I focused on two aspects of fathering journeys: mobile subjectivities as a search for identities or, alternatively, movement as an escape from identity. Both aspects are equally plausible in the same place at the same time, and perhaps a third alternative is also possible: fathering journeys as an escape from the dissonance between where I am and where I want to be, without any specific destination in mind. The journey may follow no particular route or, like Tony, it may follow a route so familiar that it requires little thought. For Tony, home is a coherent material reality, a fixed territorial entity with specific historical and geographical coordinates. What he finds today is a not aligned precisely with those coordinates and this moves him to actions such as starting a neighbourhood watch. Since the interview from which this chapter is derived, Tony has gotten involved with writing proposals for federal grants to improve the community. The money will help with everything from addressing some of the drug issues that permeate the neighbourhood to improving the common areas with landscaping. For Tony, his house and community are a materialization of his fathering, but they do not fix identity. Rather, they anchor being and create a continuity between the past, the present and a hoped-for future.

Conclusion
Fathers Beside Themselves

> To think, to write, to be, is no longer for some of us simply to follow in the tracks of those who initially expanded and explained *our* world as they established the frontiers of Europe, of Empire, and of manhood, where the knots of gendered, sexual and ethnic identity were sometimes loosened, but more usually tightened. Nor is it to echo the mimicries of ethnic absolutisms secured in the rigid nexus of tradition and community, whether in nominating our own or others' identities. It is rather to abandon such places, such centres, for the migrant's tale, the nomad's story. It is to abandon the fixed geometry of sites and roots for the unstable calculations of transit (Chambers 1994a, 246).

Becoming Father

Paul Ricoeur's idea of being and becoming is one of mutability and transference. His notion of narrative identity draws on the ways events (dwellings, fathers, children) configure into a narrative in the light of stories told by our culture, our homes, our places. We constantly reinterpret these events, becoming the "narrator of our own story without completely becoming the author of our own life" (Ricouer 1991, 437). The story is always incoherent and partial, it never has a beginning or an ending. Iris Marion Young (1997, 73) notes that "narrative provides an important way to demonstrate need." I believe that part of the need of the act of father-subject-self is to create coherency and generate meaning. As Seyla Benhabib (1999, 353) notes, becoming is about "the capacity to generate meaning over time so as to hold past, present, and future together." This book is not about fathering as an identity, nor is it about generating meaning; it is, rather, about what things are done to create the ongoing contexts of fathering. More specifically, it is about the emotional work of fathering.

The fathers in this book provide stories – in poetry, prose and conversation – that provide coherency and generate meanings that are uniquely their own. It is not my intent to suggest any consolidating themes of fatherhood writ large in contemporary Anglo-American society. If the book explores normative views of 'being a father', it does so through a number of different viewpoints, through a variety of modes of encounter. The chapters in this book move between travelling stories and connect them to variegated notions of fathering that circle around journeys, homes, places and communities. I focus on bits of stories that elaborate migrations, diasporas and transnationalisms, natural fathering, 'doing geographics', searching for the perfect house, and the construction of domesticity.

Media and historic representations play a huge role in the ways fathers (as well as mothers and children) are placed in society, and so I engage with those to the extent that they help me build a story of what fathering is supposed to be, may become, was once.

I try to pick apart aspects of our knowledge and conventional wisdom of the mythic ideals that help structure the gender and social-spatial relations of fatherhood as an institution, and to understand more fully how these get in the way of the day-to-day work of fathering. As part of this I explore how much of the institution of fatherhood hinges on an 'idea' that does not embrace the 'fact' of fathering as a daily emotional practice that is negotiated, contested, reworked and resisted differently in different spaces. That these spaces are often contrived and constructed by inchoate ideas of fatherhood foments a set of emotional practices that are awkward and incoherent.

The stories convince me that the work of fathering cultivates an important geography of care and responsibility. Beyond normative distinctions is the real world of fathering, as both *illness* and *remedy*. These are the practices, the daily work, of fathering that mirrors, placates, resists, transforms and re-invents larger societal machinations such as neo-liberalism, neo-conservatism and global economic restructuring. There are, no doubt, a number of residual aspects of essential fatherhood lurking in and around the edges of this project. It is not my concern to expose these; It is not my desire to depict the works of fathers as unconscious drives or lacks in a Lacanian or Freudian way. Nor is it my intent to uncover violent or oppressive forms of fathering lurking beneath aesthetics of care. I think they may well be there and I am happy (somewhat) to accept their presence in the knowledge that there are pieces of violence and oppression in all of us. Rather, I am drawn by Eve Sedgwick's (2003, 8) redoubtable attempt to not be caught in an attempt to get beneath and beyond her subject:

> *Beneath* and *beyond* are hard enough to let go of; what has been even more difficult is to get a distance from *beyond*, in particular the bossy gesture of 'calling for' an imminently perfect critical or revolutionary practice that one can oneself only adumbrate.

Sedgwick takes on a spatial positioning of 'beside' and, by so doing, invokes "a Deleuzian interest in planar relations," in flat ontologies that fight against any form of hierarchical and scalar power relations.

> *Beside* is an interesting preposition also because there's nothing dualistic about it: a number of elements may lie alongside one another, though not an infinity of them. *Beside* permits spacious agnosticism about several of the linear logics that enforce dualistic thinking: noncontradiction or the law of the excluded middle, cause versus effect, subject versus object. It's interest does not, however, depend on a fantasy of metonymically egalitarian or even specific relations, as any child knows who's shared a bed with siblings. *Beside* comprises a wide range

> of desiring, identifying, representing, repelling, parallelling, differentiating, rivalling, leaning, twisting, mimicking, withdrawing, attracting, aggressing, warping and other relations (Sedgwick 2003, 8).

Beside also suggests a coming together in communities that are flat and egalitarian. As the societal discourses and personal narratives progress, playfully, experimentally, and with moments of surprise (at least for me) through the pages of this project, my goal is to reach for (and not yet attain) a *coming community* that describes 'fathers beside themselves'.

Coming Communities

> To ensure human survival everywhere in the world, females and males organize themselves into communities. Communities sustain life – not nuclear families, or the 'couple,' and certainly not the rugged individual. There is no better place to learn the art of loving than in community ... We are all born into the world of community. Rarely does a child come into the world in isolation, with only one or two onlookers. Children are born into a world surrounded by the possibility of communities. Family, doctors, nurses, midwives, and even admiring strangers comprise this field of connections, some more intimate than others (hooks 2000, 129–30).

Raymond Williams (1985, 65–6) points out that unlike other terms for social relations such as state, nation or society the term 'community' is almost never used unfavourably and is often assigned an approbative sense of political unity. But the term community is problematic if it is thought of as a fusion of subjects one with another: practices of this kind exclude people who are construed as different. The desire to merge together in this kind of community generates a logic of hierarchical opposition, a separation of the pure and the impure, the inside and the outside. What I work for here is flat and relational; fathers beside themselves.

If I agree with Young (1990, 302) that the term community problematically expresses a desire to produce social wholeness and mutual identification by overcoming individualism and difference, then I want to offer a counter-topography. Doreen Massey (2005, 94) says it best when she notes that "one of the truly productive characteristics of material spatiality ... [is] its potential for the happenstance juxtaposition of previously unrelated trajectories." She goes on to elaborate that "the business of walking around a corner and bumping into alterity, of having (somehow, and well or badly) to get on with neighbours who have got 'here' ... by different routes from you; your being here together is, in that sense, quite uncoordinated. This is an aspect of the productiveness of spatiality which may enable 'something new' to happen." The push of heart work to caring and responsibility, as part of this counter-topography, is where a power over the politics of otherness finds a form that is material and geographic.

What I see in the stories that fill the pages of this book are fathers becoming something other than their fathers, other than the dictates of patriarchy, and other than the expectations of neoliberalism and neoconservatism. There are, at times and in certain places, a coming together. A sharing of emotions and practices fomented in ideals of care and responsibility.

What I give here, as best as I am able to, are the emotional connections I have had with fathers: some deep and long lasting, some fleeting. What I offer is partial and limited by my stilted abilities to write and to tell stories. The fathers' narratives – their spatial stories – are at times glorious and at times frightening. I wish it were possible to take you for a moment beyond the constraints of my writing of their worlds and into the experiences, deeply felt, of the men whose fathering stories comprise this book.

I am surrounded by men and women who help me with my fathering. We meet for coffee, for breakfast, for a walk along the beach with our dogs. We meet in pairs and in large groups. Sometimes I shun the support and, even at those times, I know it is there and available. This is my coming community. I do not know its form and I do not understand how it works. The messiness is important.

In *The Inoperative Community*, Jean-Luc Nancy (1991) notes that communities cannot be built upon already constituted subjects. There must be surprise from throwntogetherness, movements towards acceptance of the happenstance. Subjects are never fully constituted, they are always becoming. For Nancy, community is not about common identities and common being, but rather it is about "being *in* common" (Nancy 1991, 4). This is a community that is *becoming other* in as yet unthought of ways.

My coming community is a practice that began before my children Ross and Catherine were born, and will continue in a different way now that my son has moved out of our home and on to university, and my daughter begins her search for what to do after high school. I am filled with joy and grief at the prospect of their moving on. Out for breakfast with some of my male friends the other morning I found myself sharing some of the moments I have spent with my son and daughter over the last few years: the serenity of paddling a double kayak, the tears at the side of a swimming pool, the pride of a high school graduation, the quiet repose of a walk along the beach, the tired satisfaction of back-packing in the Sierras, the quiet of a shared breakfast burrito, the exhilaration of rafting down a raging river. The memories of these events are realized in tears, laughter, sadness and joy. This is the emotional work of fathering.

As I review this final chapter I am sitting beside a swimming pool watching Catherine compete in the 400m free-style of San Diego County's Junior Olympics. She is good at distance, with the mile as her favourite. It has been a tough year for her because of a shoulder injury. She is coming through in this race and will win a medal. As I watch her, like many times before, my eyes start glassing over. It is not about winning medals or about parental pride, it is about the ways I see her grow into the young woman she is becoming, the community of friends she has created with her swim team, and her capacity for joy and acceptance.

Three years ago, an old friend invited my son and I on a canoe trip down the Yukon River. At the time he was a priest in Dawson City and he wanted to share the magnificence of a small part of Canada's North-West territories with some of his friends. Originally advertised as a father-son trip, Ross and I were the only father and son out of the 12 participants, of whom I knew about half. At 15 years of age, Ross was the youngest on the trip. At the end of the trip – before the first of our crew had to leave to catch a flight – we circled up to share our feelings. This is a practice we'd gotten into every morning before getting into the canoes: we circled on the bank of the river and shared our joy, fear, pain or misery of the day. On this last day we circled up on the grass in front of the priest's house. Each man was to share the one most important thing that he got out of the trip. As each man shared, the circle heard again about the magnificence of the hills we passed through, the fear of grizzly bears, the speed of the river, the island refuges, the wonder of the bald eagles, the misery of the rain, the anxiety when we lost touch with two men and their canoe for a short period of time, the beauty of the last turn into Dawson. My son was next to last to share. He loves nature and so my expectation was he'd share something about the wildness of the trip. He said simply, "the most important thing to me was these circles and you guys sharing from your hearts." At that moment I lost any semblance of composure; my body heaved with inconsolable sobs. I was the last to share and I could not speak. One of my friends noted "Well, I think Stuart is sharing his most important truth." Each day I work to create the kind of safe place that enables this kind of sharing because, by so doing, community comes.

References

Addelson, K (2005), Foresight and Memory. Unpublished book manuscript. Leverett, MA (cited in Gibson-Graham 2006).

Adorno, Theodore (1974), Refuge for the Homeless. Translated by E.F.N. Jephcott, *Minima Moralia: Reflections from a Damaged Life*, pp. 38–9 (London: New Left Books).

Agamben, Giorgio (1993), *The Coming Community.* Translated by Michael Hardt (London and Minneapolis: University of Minnesota Press).

Aitken, Stuart C. (1994), *Putting Children in Their Place* (Washington DC: Association of American Geographers Resource Publication Series and Boston: Edwards Bros).

______ (1998), *Family Fantasies and Community Space* (New Brunswick, NJ: Rutgers University Press).

______ (1999), Putting Parents in Their Place: Child Rearing Rites and Gender Politics. In Elizabeth K. Teather (ed.) *Geographies of Personal Discovery: Places, Bodies and Rites of Passage*, pp. 104–25 (London: Routledge).

______ (2000a), Play, Rights and Borders: Gender Bound Parents and the Social Construction of Children. In Sarah Holloway and Gill Valentine (eds) *Children's Geographies: Living, Playing, Learning and Transforming Everyday Worlds*, pp. 119–38 (London: Routledge).

______ (2000b), Fear, Loathing and Space for Children. In John R. Gold and George Revill (eds) *Landscapes of Defense*, pp. 48–67 (Harlow, UK: Prentice Hall).

______ (2000c), Fathering and Faltering: "Sorry, But You Don't Have the Necessary Accoutrements". *Environment and Planning A*, 32(4), 581–98.

______ (2000d), Mothers, Communities and the Scale of Difference. *Journal of Social and Cultural Geography*, 1(1), 69–86.

______ (2001a), Shared Lives. In Melanie Limb and Claire Dwyer (eds) *Qualitative Methods for Geographers*, pp. 73–86 (New York and London: Arnold Publishers).

______ (2001b), *Geographies of Young People: The Morally Contested Spaces of Identity* (London and New York: Routledge).

______ (2005), The Awkward Spaces of Fathering. In Bettina van Hoven and Kathrin Hoerschelmann (eds) *Spaces of Masculinity*, pp. 222–37 (New York and London: Routledge).

______ (2006), Leading Men to Violence and Creating Spaces for their Emotions. *Gender, Place and Culture*, 13(5), 491–507.

______ (2007), *Desarrollo Integral y Fronteras*/Integral Development and Borderspaces. *Children's Geographies*, 5(1–2), 113–29.

Aitken, Stuart C. and Deborah Dixon (forthcoming), Avarice and Tenderness in Cinematic Landscapes of the American West. In Dydia Delyser and Michael Dear (eds) *Geography and the Humanities.*

Aitken, Stuart C. and Thomas Herman (1997) Gender, Power and Crib Geography: From Transitional Spaces to Potential Places. *Gender, Place and Culture: A Journal of Feminist Geography*, 4(1), 63–88.

Aitken, Stuart C. and Chris Lukinbeal (1997), Mobility, Road Geographies and the Quagmire of Terra Infirma. In Steven Cohen and Ina Rae Hark (eds) *Road Movies*, pp. 349–70 (London: Routledge).

______ (1998), Of Heroes, Fools and Fisher Kings: Cinematic Representations of Street Myths and Hysterical Males. In Nick Fyfe (ed.) *Images of the Street*, pp. 141–59 (London: Routledge).

Aitken, Stuart C. and Leo E. Zonn (1994), *Power, Place, Situation and Spectacle: A Geography of Film* (Lanham, MD: Rowman and Littlefield).

Alanen, Leena (1994), Gender and Generation: Feminism and the "Child Question." In J. Qvotrup, M. Brady, G. Sgritta and H. Winterberger (eds) *Childhood Matters: Social Theory, Practice and Politics*, pp. 27–42 (Aldershot, UK: Avebury Press).

______ (2001), Explorations in Generational Analysis. In L. Alanen and B. Mayall (eds) *Conceptualizing Child-Adult Relations*, pp. 11–22 (London and New York: Routledge).

Anderson, Benedict (1991), *Imagined Communities* (London: Verso).

Aoyama, Tomoko (2007), Father-Daughter Relationships. In M. Flood, J. Kegan-Gardiner, B. Pease and K. Pringle (eds) *International Encyclopedia of Men and Masculinities*, pp. 189–90 (London and New York: Routledge).

Bachelard, Gaston (1964), *The Poetics of Space*. Translated by Maria Jolas (Boston: Beacon Press).

Bammer, A. (1992), Editorial. *New Formations*, 17, vii-xi.

Barnes, Trevor and Jim Duncan (1992), *Writing Culture* (New York and London: Routledge).

Basu, Paul (2001), Hunting Down Home: Reflections on Homeland and the Search for Identity in the Scottish Diaspora. In Barbara Bender and Margot Winer (eds) *Contested Landscapes: Movement, Exile and Place*, pp. 333–48 (Oxford: Berg).

Baudrillard, Jean (1988), *America* (London: Verso).

Benhabib, Seyla (1999), Citizens, Residents, and Aliens in a Changing World: Political Membership in the Global Era. *Social Research*, 66(3), 352–67.

Benjamin, Jessica (1986), A Desire of One's Own: Psychoanalytic Feminism and Intersubjective Space. In Teresa de Lauretis (ed.) *Feminist Studies/Critical Studies*, pp. 78–101 (Bloomington: Indiana University Press).

Benton, Lisa (1995).Will the Real/Reel Los Angeles Please Stand Up? *Urban Geography*, 16(2), 144–64.

Bergson, Henri (1888/1988). *Matter and Memory*. Translated by N. M. Paul and W. S. Palmer (New York: Zone Books).

______ (1910), *Time and Free Will* (London: George Allen and Unwin).

Bernardes, Jon (1985), Family Ideology: Identification and Exploration. *Sociological Review*, 33, 275–97.

Bianchi, Suzanne M., John P. Robinson and Melissa A. Milkie (2006), *Changing Rhythms of American Family Life* (New York: Russell Sage Foundation).

Blakenhorn, David (1995), *Fatherless America: Confronting Our Most Urgent Social Problem* (New York: Harper Perennial).

Blunt, Allison (2006), *Domicile and Diaspora: Anglo-Indian Women and the Spatial Politics of Home* (Oxford: Blackwell).

Blunt, Allison and Ann Varley (2004), Geographies of Home. *Cultural Geographies*, 11, 3–6.

Bly, Robert (1990), *Iron John: A Book About Men* (Reading, MA: Addison-Welsey Publishing Company, Inc.).

Bondi, Liz (2005), The Place of Emotions in Research: From Partitioning Emotion and Reason to the Emotional Dynamics of Research Relationships. In Joyce Davidson, Liz Bondi and Mick Smith (eds) *Emotional Geographies*, pp. 231–46 (Aldershot: Ashgate Publishing).

Bowlby, Rachel (1995), Domestication. In Diane Elam and Robyn Wiegman (eds) *Feminism Beside Itself*, pp. 71–92 (New York: Routledge).

Brandth, B. and E. Kvande (1998), Masculinity and Child-care: The Reconstruction of Fathering. *The Sociological Review*, 46(2), 293–313.

Brownlie, Julie and Simon Anderson (2006), "Beyond Anti-Smacking:" Re-Thinking Parent-Child Relations. *Childhood*, 13(4), 479–98.

Bruno, Guilliana (2002), *The Atlas of Emotion: Journeys in Art, Architecture, and Film* (Verso: London).

Buchanan, Ian (2004), Introduction: Deleuze and Music. In Ian Buchanan and Marcel Swiboda (eds) *Deleuze and Music*, pp. 1–19 (Edinburgh: Edinburgh University Press).

Butler, Judith (1990), *Gender Trouble: Feminism and the Subversion of Identity* (London and New York: Routledge).

______ (1997), *The Psychic Life of Power* (Stanford, CA: Stanford University Press).

Butler, Ruth (1999), The Body. In Paul Cloke, Phil Crang and Mark Goodwin (eds) *Introducing Human Geographies*, pp. 238–46 (London: Arnold).

Callard, Felicity (1998), The Body in Theory. *Society and Space*, 1, 387–400.

Carrier, R.M. (1989), The Ontological Significance of Deleuze and Guattari's Concept of the Body Without Organs. *Journal of the British Society for Phenomenology*, 29, 189–206.

Chamberlain, Mary and Selma Leydesdorff (2004), Transnational Families: Memories and Narratives. *Global Networks*, 4(3), 227–41.

Chambers, Iain (1994a), Leaky Habitats and Broken Grammar. In George Robertson, Melinda Mash, Lisa Tickner, Jon Bird Barry Curtis and Tim

Putnam (eds) *Travellers' Tales: Narratives of Home and Displacement*, pp. 245–9 (London and New York: Routledge).
______ (1994b), *Migrancy, Culture, Identity* (London and New York: Routledge).
______ (2001), *Culture after Humanism: History, Culture, Subjectivity* (London: Routledge).
Chapman, T. and J. Hockey (eds, 1999), *Ideal Homes: Social Change and Domestic Life* (London: Routledge).
Cieradd, Irene (1999), *At Home: An Anthropology of Domestic Space* (Syracuse: Syracuse University Press).
Cloke, Paul and Owain Jones (2001), Dwelling, Place and Landscape: An Orchard in Somerset. *Environment and Planning A*, 33, 649–66.
Clover, Carol (1989), Her Body, Himself: Gender and the Slasher Film. In J. Donald (ed.) *Fantasy and the Cinema*, pp. 91–133 (London: British Film Institute).
Coltrane, Scott (1996), *Family Man: Fatherhood, Housework, and Gender Equity* (New York: Oxford University Press).
Conley, Tom (2007), *Cartographic Cinema* (Minneapolis: University of Minnesota Press).
Connell, Robert W. (1995), *Masculinities* (Berkeley, CA: University of California Press).
Connolly, William (1999), *Why I am Not A Secularist* (Minneapolis: University of Minnesota Press).
______ (2002), *Neuropolitics: Thinking, Culture, Speed* (Minneapolis: University of Minnesota Press).
Coontz, Stephanie (1992), *The Way We Never Were: American Families and the Nostalgia Trap* (New York: Basic Books).
Crary, David (2008), Increasingly, Men Pitch in on Housework. In *The San Diego Union-Tribune*, Thursday, March 6, A-1 and A-7.
Creswell, Tim (2004). *Place: A Short Introduction* (Oxford, UK: Blackwell Publishers).
Curry, Michael R. (1996), *The Work in the World: Geographical Practice and the Written Word* (Minneapolis: University of Minnesota Press).
Curti, Giorgio Hadi (2008), Beating Words to Life: Subtitles, Assemblage(S)capes, Expressions. *Geojournal*, 12(3), DOI: 10.1007/s10708-008-9221-1.
Debord, Guy (1983/2000), *Society of the Spectacle* (Detroit: Black and Red).
de Certeau, Michel (1984), *The Practice of Everyday Life* (Berkeley, CA: University of California Press).
______ (1988), *The Writing of History*. Translated by Tom Conley (New York: Columbia University Press).
de Certeau, Michel, Luce Giard and Pierre Mayol (1998), *The Practice of Everyday Life, Vol. 2: Living and Cooking*. Translated by Timothy J. Tomasik (Minneapolis: University of Minnesota Press).
Del Castillo, Adelaida (1993), Covert Cultural Norms and Sex/Gender Meaning: A Mexico City Case. *Urban Anthropology*, 22(3–4), 237–58.

Deleuze, Gilles (1983), *Nietzsche and Philosophy* (New York: Columbia University Press).

______ (1986), *Foucault*. Translated and edited by S. Hand (Minneapolis: University of Minnesota Press).

______ (1988), *Spinoza: Practical Philosophy* (San Francisco: City Light Books).

______ (1989), *Cinema 2: The Time-Image* (Minneapolis: University of Minnesota Press).

______ (1990), *Expressionism in Philosophy: Spinoza* (New York: Zone Books).

______ (1994), *Difference and Repetition* (New York: Columbia University Press).

______ (2001), *Pure Immanence: Essays on A Life*. Translated by John Rajchman (New York: Zone Books).

______ (2005), *Expressionism in Philosophy: Spinoza* (New York: Zone Books).

Deleuze, Gilles and Félix Guattari (1983), *Anti-Oedipus: Capitalism and Schizophrenia* (Minneapolis: University of Minnesota Press).

______ (1987), *A Thousand Plateaus: Capitalism and Schizophrenia* (London: The Athlone Press).

Derrida, Jacques (2001), *On Cosmopolitanism and Forgiveness* (London: Routledge).

______ (2002), Derelictions of the Right to Justice (but What are the 'Sans-Papiers' Lacking?). In Elizabeth Rottenberg (ed.) *Negotiations: Interventions and Interviews 1971–2001*, pp. 133–44 (Stanford, CA: Stanford University Press).

Dienhart, Anna (1998), *Reshaping Fatherhood: The Social Construction of Shared Parenting* (Thousand Oaks and London: Sage).

Doel, Marcus (1999), *Poststructural Geographies: The Diabolical Art of Spatial Science* (Edinburgh: Edinburgh University Press).

Doherty, W.J., E.F. Kouneski and M.F. Erickson (1998), Responsible Fathering: An Overview and Conceptual Framework. *Journal of Marriage and the Family*, 60, 277–92.

Due, Reidar (2007), *Deleuze*. Key Contemporary Thinkers Series (London: Polity Press).

Dwyer, Chris (2002), "Where are You From?" Young British Muslim Women and the Making of Home. In A. Blunt and C. McEwan (eds) *Postcolonial Geographies*, pp. 184–99 (London: Continuum).

Dyck, Isabel (1990), Space, Time and Renegotiating Motherhood: An Exploration of the Domestic Workplace. *Society and Space*, 8, 459–83.

Eisenstein, Zillah R. (1981), *The Radical Future of Liberal Feminism* (New York: Longman).

Elden, Stuart (2001), *Mapping the Present: Heidegger, Foucault and the Project of Spatial History* (London: Continuum).

Elshtain, Jean B. (1990), The Family in Political Thought: Democratic Politics and the Question of Authority. In J. Sprey (ed.) *Fashioning Family Theory*, pp. 51–66 (Newbury Park, CA: Sage).

England, Kim (ed., 1996), *Who Will Mind the Baby? Geographies of Child Care and Working Mothers* (London and New York: Routledge).

Enloe, Cynthia (1990), "Just like One of the Family": Domestic Servants in World Politics. In C. Enloe *Bananas, Beaches and Bases: Making Feminist Sense of International Politics*, pp. 177–94 (Berkeley: University of California Press).

Faludi, Susan (1999), *Stiffed: The Betrayal of American Men* (New York: William Morrow and Company, Inc.).

Filmer, Robert (1680), *Patriarcha, Or, the Natural Power of Kings* (Printed for Ric. Chiswell, Matthew Gillyflower and William Henchman).

Foord, Jo and Nicky Gregson (1986), Patriarchy: Towards a Reconceptualization. *Antipode*, 18, 181–211.

Foster, David (1994), Taming the Father: John Locke's Critique of Patriarchal Fatherhood. *The Review of Politics*, 56(4), 641–67.

Foucault, Michel (1977), *Discipline and Punish: The Birth of the Prison* (New York: Vintage Books).

Fraad, H., S. Resnick and R. Wolff (1994), *Bringing It All Back Home: Class, Gender, and Power in the Modern Household* (London: Pluto Books).

French, Marilyn (1971), *The Women's Room* (New York: Jove Publications).

Gadd, David (2003), Reading Between the Lines: Subjectivity and Men's Violence. *Men and Masculinities*, 5(4), 333–54.

George, Rosemary Marangoly (1998a), *Burning Down the House: Recycling Domesticity* (Boulder, CO: Westview Press).

______ (1998b), Recycling: Routes to and From Domestic Fixes. In Rosemary Marangoly George (ed.) *Burning Down the House: Recycling Domesticity*, pp. 1–22 (Boulder, CO: Westview Press).

______ (1998c), Homes in the Empire, Empire in the Home. In Rosemary Marangoly George (ed.) *Burning Down the House: Recycling Domesticity*, pp. 23–46 (Boulder, CO: Westview Press).

Gibson-Graham, J.K. (1996), *The End of Capitalism (As We Knew It)* (Minneapolis: University of Minnesota Press).

______ (2006), *A Postcapitalist Politics* (Minneapolis: University of Minnesota Press).

Gilson, Etienne (1964), Foreword to Gaston Bachelard's *The Poetics of Space* (Boston: Beacon Press).

Glenn, Evelyn Nakano (1987), Gender and the Family. In Beth Hess and Myra Marx Feree (eds) *A Handbook of Social Science Research*, pp. 348–80 (Newbury Park, CA: Sage).

González-López, Gloria (2007), Hombre. In M. Flood, J. Kegan-Gardiner, B. Pease and K. Pringle (eds) *International Encyclopedia of Men and Masculinities*, p. 306 (London and New York: Routledge).

Guthman, Edward (2000), Stunning 'Beauty' from Mendes. *San Francisco Chronicle*, May 10, C-17.

Guttmann, Matthew C. (1996), *The Meaning of Macho* (Berkeley, CA: University of California Press).

______ (ed., 2003), *Changing Men and Masculinities in Latin America* (Durham, NC: Duke University Press).

______ (2007), Machismo (and Macho). In M. Flood, J. Kegan-Gardiner, B. Pease and K. Pringle (eds) *International Encyclopedia of Men and Masculinities*, p. 372 (London and New York: Routledge).

Hall, Stuart (1990), Cultural Identity and Diaspora. In J. Rutherford (ed.) *Identity: Community, Culture, Difference*, pp. 222–37 (London: Lawrence and Wishart).

Hansen, Karen Tranberg (1992), *African Encounters with Domesticity* (New Brunswick, NJ: Rutgers University Press).

Haraway, Dona (1991), *Simians, Cyborgs, and Women: The Reinvention of Nature* (New York and London: Sage).

Harker, Christopher (2005), Playing and Affective Time-Spaces. *Children's Geographies*, 3(1), 47–62.

Harrison, Paul (2007), The Space Between Us: Opening Remarks on the Concept of Dwelling. *Environment and Planning D: Society and Space*, 25, 625–47.

Harvey, David (2000), *Spaces of Hope* (Berkeley, CA: University of California Press).

Hearn, Jeff (1996), Is Masculinity Dead? A Critique of the Concept of Masculine/Masculintities. In Máirtín Mac an Ghaill (ed.) *Understanding Masculinities: Social Relations and Cultural Arenas*, pp. 202–17 (Buckingham and Philadelphia: Open University Press).

Heidegger, Martin (1993, 1951), Dwelling Building Thinking. In D. Farell Krell (ed.) *Basic Writings*, pp. 347–63 (London: Routledge).

Herman, Arthur (2001), *How the Scots Invented the Modern World* (New York: Three Rivers Press).

Hochschild, Arlie (1989), *The Second Shift: Working Parents and the Revolution at Home* (New York: Viking).

Holloway, Sarah (1998), Local Childcare Cultures: Moral Geographies of Mothering and the Social Organization of Pre-School Education. *Gender, Place and Culture*, 5(1), 29–53.

Honig, Bonnie (1994), Difference, Dilemmas and the Politics of Home. *Social Research*, 61, 563–97.

hooks, bell (1990), *Yearning: Race, Gender, and Cultural Politics* (Boston: South End Press).

______ (1996), *Reel to Real: Race, Sex, and Class at the Movies* (New York and London: Routledge).

______ (2000), *All About Love: New Visions* (New York: Perennial).

Hume, David (1739/1955), *A Treatise on Human Nature: Being an Attempt to Introduce the Experimental Method of Reasoning in Moral Subjects.* Edited with an analytical text by L.A. Selby-Bigge (London: Clarendon Press).

Hyams, Melissa (2003), Adolescent Latina Bodyspaces: Making Homegirls, Homebodies and Homeplaces. In Katharyne Mitchell, Sallie Marston and

Cindi Katz (eds) *Life's Work: Geographies of Social Reproduction*, pp. 119–40 (Oxford: Blackwell).
Jackson, Peter and Polly Russell (forthcoming), Life History Interviewing. In Dydia Delyser, Steve Herbert, Stuart Aitken, Mike Crane and Linda McDowel (eds) *The Handbook of Qualitative Research in Geography* (London: Sage).
Kaika, Maria (2004), Interrogating the Geographies of the Familiar: Domesticating Nature and Constructing the Autonomy of the Modern Home. *International Journal of Urban and Regional Research*, 28, 265–86.
Kandiyotti, Deniz (1988), Bargaining with Patriarchy. *Gender and Society*, 2, 274–90.
Kaplam, Caren (1987), Deterritorializations: The Rewriting of Home and Exile in Western Feminist Discourse. *Cultural Critique*, 6, 187–98.
Kibbe, Pauline R. (1946), *Latin Americans in Texas* (Albuquerque: University of New Mexico Press).
Knopp, Larry (2003), Movement and Encounter. In Stuart C. Aitken and Gill Valentine (eds) *Approaches to Human Geography*, pp. 218–25 (London, Thousand Oaks, CA, and New Delhi: Sage).
Knox, Paul (1991), The Restless Urban Landscape: Economic and Sociocultural Change in the Transformation of Washington DC. *Annals of the Association of American Geographers*, 81(2), 181–209.
Kobayashi, Audrey (1994), Coloring the Field: Gender, "Race," and the Politics of Fieldwork. *The Professional Geographer*, 46(1), 73–80.
Kolmerten, Carol A. (1990), *Women in Utopia: The Ideology of Gender in the American Owenite Communities* (Bloomington, IN: Indiana University Press).
Kristeva, Julie (1983), *Power of Horrors* (New York: Columbia University Press).
______ (1986), *The Kristeva Reader*. Edited by T. Moi (New York: Columbia University Press).
Lacan, Jacques (1978), *The Four Fundamental Concepts of Psychoanalysis* (New York: Verso).
Laclau, Ernesto (1990), *New Reflections on the Revolution of Our Time* (London: Verso).
Laclau, Ernesto and Chantal Mouffe (2001), *Hegemony and Socialist Strategy*, 2nd edition (London: Verso).
Lamb, Michael (ed., 1997), *The Role of the Father in Child Development*, 3rd Edition (New York: John Wiley and Sons).
Laplanche, J. and J.B. Pontalis (1972), *The Language of Psychoanalysis*. Translated by D. Nicholson-Smith (New York: Norton).
______ (1986), Fantasy and the Origins of Sexuality. In V. Burgin, J. Donald and C. Kaplan (eds) *Formations of Fantasy*, pp. 5–34 (Boston: Harvard University Press).
Laurie, Nina and Liz Bondi (eds, 2005), *Working the Spaces of Neoliberalism* (Oxford: Blackwell).

Laqueur, Thomas W. (1990), *Making Sex: Body and Gender from the Greeks to Freud* (Cambridge, MA: Harvard University Press).

______ (1992), The Facts of Fatherhood. In Barrie Thorne and Marilyn Yalom (eds) *Rethinking the Family: Some Feminist Questions*, pp. 140–54 (Boston: Northeastern University Press).

Lawrence, D.H. (1913/1976), *Sons and Lovers* (Dallas: Penguin Books).

Lea, Michael (2008), Fight for Fatherhood. *Daily Mail*, Monday, May 19, p. 10.

LeFebvre, Henri (1991), *The Production of Space*. Translated by Donald Nicholson-Smith (Oxford: Blackwell).

LePlay, Frédéric (1871/1982), *On Family, Work and Social Change* (Chicago: University of Chicago Press).

Levinas, Emmanuel (1969), *Totality and Infinity: Essays on Exteriority*. Translated by A. Lingus (Indianapolis: Indiana University Press).

Locke, John (1690/1824), *Two Treatises on Government* (London: C. and J. Riverton et al., and Edinburgh: Stirling and Slade).

______ (1693), *Thoughts Concerning Education* (London: A. and J. Churchill at the Black Swan in Paternoster-row).

Lombardo, Paul and Gregory Dorr (2006), Eugenics, Medical Education, and the Public Health Service: Another Perspective on the Tuskegee Syphilis Experiment. *Bulletin of the History of Medicine*, 80, 291–316.

Lukinbeal, Chris and Stuart C. Aitken (1998), Sex, Violence and the Weather: Male Hysteria, Scale and the Fractal Geographies of Patriarchy. In Steve Pile and Heidi Nast (eds) *Places Through the Body*, pp. 356–80 (London: Routledge).

Lulka, David (2004), Stablizing the Herd: Fixing the Identity of Nonhumans. *Society and Space*, 22(3), 439–463.

Lupton, Deborah and Lesley Barclay (1997), *Constructing Fatherhood: Discourses and Experiences* (Thousand Oaks and London: Sage).

Mackenzie, Suzanne and Damaris Rose (1983), Industrial Change, the Domestic Economy and Home Life. In J. Anderson, S. Duncan and R. Hudson (eds) *Redundant Spaces in Cities and Regions: Studies in Industrial Decline City Change*, pp. 155–200 (London: Academic Press).

Malinowski, B. (1913), *The Family Among the Australian Aborigines* (London: University of London Press).

Malpas, J.E. (1999), *Place and Experience: A Philosophical Topography* (Cambridge, UK: Cambridge University Press).

Manning, Wendy and Pamela Smock (2005), Measuring and Modeling Cohabitation: New Perspectives from Qualitative Data. *Journal of Marriage and the Family*, 67(4), 989–1002.

Marston, Sallie, John Paul Jones and Keith Woodward (2005), Human Geographies Without Scale. *Transactions of the British Institute of Geography*, 30(4), 416–32.

Marston, Sallie and Katharyne Mitchell (2004), Citizens and the State: Citizen Formations in Space and Time. In Clive Barnett and Murray Low (eds) *Space*

of Democracy: Geographical Perspectives on Citizenship, Participation and Representation, pp. 93–112 (London: Sage).
Massey, Doreen (2005), *For Space* (London: Sage).
Massumi, Brian (1992), *A User's Guide to Capitalism and Schizophrenia* (Cambridge, MA: MIT Press).
______ (2002), *Parables for the Virtual: Movement, Affect, Sensation* (London: Duke University Press).
McCann, Eugene (1995), Neotraditional Developments: The Anatomy of a New Urban Form. *Urban Geography*, 16(3), 210–233.
McClintock, Ann (1995), *Imperial Leather: Race, Gender, and Sexuality in the Colonial Contest* (London and New York: Routledge).
McCormack, Derek P. (2003), An Event of Geographical Ethics in Spaces of Affect. *Transactions of the Institute of British Geographers*, 28(4), 488–507.
McKenzie, Evan (1994), *Privatopia: Homeowners Associations and the Rise of Residential Private Government* (New Haven, CT: Yale University Press).
McKinney, Irene (1989), Review of McKinney's *Six O'Clock Mine Report* (Pittsburgh: University of Pittsburgh Press). http://www.wvwc.edu/lib/wv_authors/authors/a_mckinney.htm.
Miller, D. (ed., 2001), *Home Possessions: Material Culture Behind Closed Doors* (Oxford: Berg).
Minh-ha, Trinh (1994), Other than Myself/My Other Self. In George Robertson, Melinda Mash, Lisa Tickner, Jon Bird, Bary Curtis and Tim Putman (eds) *Travellers' Tales: Narratives of Home Displacement*, pp. 9–28 (London and New York: Routledge).
Mintz, Steve (1997), From Patriarchy to Androgyny and Other Myths: Placing Men's Family Roles in Historical Perspective. In Alan Booth and Ann C. Crouter (eds) *Men in Families*, pp. 3–30 (Mahwah, NJ: Lawrence Erlbaum Associates).
Mitchell, Katharyne, Sallie Marston and Cindi Katz (eds) (2004), *Life's Work: Geographies of Social Reproduction* (Oxford: London).
Mitchell, Lee Clark (1996), *Westerns: Making the Man in Fiction and Film* (Chicago and London: University of Chicago Press).
Moosa-Mitha, Mehmoona (2005), A Difference-Centred Alternative to Theorization of Children's Citizenship Rights. *Citizenship Studies*, 9(4), 369–88.
More, Thomas (1715/1885), *Utopia.* Translated by Gilbert Burnet (London: George Routledge and Sons).
Morris, Meaghan (1992), Great Moments in Social Climbing: King Kong and the Human Fly. In B. Colomina (ed.) *Sexuality and Space*, pp. 1–51 (Princeton, NJ: Princeton Architectural Press).
Mouffe, Chantal (1992), Democratic Citizenship and the Political Community. In the Miami Theory Collective, *Community at Loose Ends* (Minneapolis: University of Minnesota Press).
______ (1995), Post-Marxism: Democracy and Identity. *Environment and Planning D: Society and Space*, 13, 259–65.

Murdock, George Peter (1949), *Social Structure* (New York: Free Press).
Nancy, Jean Luc (1991), *The Inoperative Community* (Minneapolis: University of Minnesota Press).
Nelson, Craig (2006), *Thomas Paine: Enlightenment, Revolution and the Birth of Modern Nations* (New York and London: Penguin).
Nelson, Lise (1999), Bodies (and Spaces) Do Matter: The Limits of Performativity. *Gender, Place and Culture*, 6(4), 331–53.
Ngai, Mae (2004), *Impossible Subjects: Illegal Aliens and the Making of Modern America* (Princeton, NJ: Princeton University Press).
Owen, Robert (1816/1972), *A New View of Society* (London: Macmillan Press Ltd).
Owen, U. (ed., 1983), *Fathers: Reflections by Daughters* (London: Virago).
Paine, Thomas (1776/1979), *Common Sense, The American Crisis, The Age of Reason*, Limited Edition (Pennsylvania: The Franklin Library).
______ (1783/1970), *The Rights of Man and Other Writings* (London: Heron Books).
Palmer, Phyllis (1989), *Domesticity and Dirt: Housewives and Domestic Servants in the US, 1920–1945* (Philadelphia: Temple University Press).
Peck, M. Scott (1993), *Further Along the Road Less Traveled: The Unending Journey Towards Spiritual Growth* (New York: Touchstone Press).
Pile, Steven (1996), *The Body and the City* (London and New York: Routledge).
Pleck, E.H. and J.H. Pleck (1997), Fatherhood Ideals in the United States: Historical Dimensions. In M. Lamb (ed.) *The Role of Father in Child Development*, 3rd Edition, pp. 33–48 (New York: John Wiley).
Popenoe, David (1993), American Family Decline, 1960–1990: A Review and Appraisal. *Journal of Marriage and Family*, 55, 527–55.
______ (1996), *Life Without Father: Compelling New Evidence that Fatherhood and Marriage are Indispensible for the Good of Children and Society* (Cambridge, MA: Harvard University Press).
Pratt, Geraldine (1998), Geographical Metaphors in Feminist Theory. In Susan Aitken, Ann Brigham, Sallie Marston and Pennv Waterstone (eds) *Making Worlds: Gender,. Metaphor, Materiality*, pp. 13–30 (Tucson, AZ: University of Arizona Press).
______ (2004), Valuing Childcare: Troubles in Suburbia. In Katharyne Mitchell, Sallie Marston and Cindi Katz (eds) *Life's Work: Geographies of Social Reproduction*, pp. 164–84 (Oxford: Blackwell).
Pryor, C. (1999), Welcome Back to the Family. *The Australian*, May 19, p. 14.
Pryor, Jan and Bryan Rodgers (2001), *Children in Changing Families: Life After Parental Separation* (Oxford: Blackwell).
Rahman, Najat (2007), Patriarchy. In M. Flood, J. Kegan-Gardiner, B. Pease and K. Pringle (eds) *International Encyclopedia of Men and Masculinities*, pp. 468–70 (London and New York: Routledge).
Rich, Adrienne (1977), *Of Woman Born: Motherhood as Experience and Institution* (New York: Norton).

Riessman, C.K. (1990), Strategic Uses of Narrative. *Social Science and Medicine*, 30, 1195–200.
______ (1993), *Narrative analysis* (London: Sage).
Ritter, Alan and Julia Conway Bondnella (eds, 1988), *Rousseau's Political Writings* (New York: Norton Critical Editions).
Rivoeur, Paul (1991), Life: A Story in Search of a Narrator. In M. Valdes (ed.) *A Ricoeur Reader: Reflection and Imagination*, pp. 425–37 (Hemel Hempstead, Herts: Harvester Wheatsheaf).
Roberts, Joanna (2005), *Telling the Tale: Narratives of Place, Masculinity and Health in Foxhill-Parson Cross, Sheffield*. Unpublished PhD thesis, University of Sheffield (cited in Peter Jackson and Polly Russell, forthcoming).
Roberts, Marion (1990), Gender and Housing. *Built Environment*, 16(4), 257–68.
Rousseau, Jean-Jacques (1755/1988), Discourses on Political Economy. Translated by J.C. Bondanella. In A. Ritter and J.C. Bondanella (eds) *Rousseau's Political Writings*, p. 64 (New York: Norton).
______ (1761/1987), *La Novelle Héloise* (College Station, PA: Pennsylvania State University Press).
______ (1762/1962), *The* Émile *of Jean-Jacques Rousseau: Selections*. Translated and edited by William Boyd (New York: Columbia University, Bureau of Publications).
______ (1763/1901), Rousseau's *Social Contract*. In *Ideal Empires and Republics*, pp. 1–128 (New York and London: M. Walter Dunne).
Ruddick, Sara (1992), Thinking about Fathers. In Barrie Thorne and Marilyn Yalom (eds) *Rethinking the Family: Some Feminist Questions*, pp. 176–90 (Boston: Northeastern University Press).
Saetz, Stephen B., Marian Van Court and Mark W. Henshaw (1985), Eugenics and the Third Reich. *The Eugenics Bulletin*. Reprinted on Future Generations website at http://www.ziplink.net/~bright/.
Sartre, Jean-Paul (no date), *Being and Nothingness*. Translated with an introduction by Hazel E. Barnes (New York: Philosophical Library).
Sauer, M. (1993), Chore War: Which Partner is Doing What? How Willingly? *The San Diego Union-Tribune*, September 25, Sec. E. p. 1.
Secunda, Vitoria (1992), *Women and their Fathers: The Sexual and Romantic Impact Of The First Man In Your Life* (New York: Doubleday Dell Publishing Group).
Sedgwick, Eve Kosofsky (2003) *Touching Feeling: Affect, Pedagogy, Performativity* (Durham and London: Duke University Press).
Semetsky, I. (2004), Becoming-Language/Becoming-Other: Whence Ethics? *Educational Philosophy and Theory*, 36(3), 313–25.
Sibley, David (2005), Private/Public. In David Atkinson, Peter Jackson, David Sibley and Nigel Washbourne (eds) *Cultural Geography: A Critical Dictionary of Key Concepts*, pp. 155–60 (London: I B Taurus).
Silverman, Kaja (1992), *Male Subjectivity at the Margins* (New York and London: Routledge).

Simonsen, Kirsten (2000), Editorial: The Body as Battlefield. *Transactions of the Institute of British Geographers*, 25, 7–9.

Sinclair, Upton (1927), *Oil!* (New York: Penguin Books).

Smith, David L. (2002), 'Beautiful Necessities': *American Beauty* and the Idea of Freedom. *Journal of Religion and Film*, 6(2), http://www.unomaha.edu/jrf/am.beauty.htm.

Smith, Neil (1993), Homeless/Global: Scaling Places. In Jon Bird, Bary Curtis, Tim Putman, George Robertson and Lisa Tickner (eds) *Mapping the Futures: Local Cultures, Global Change*, pp. 87–119 (London: Routledge).

St. George, Donna (2007), Study: June Cleaver has Nothing on Today's Moms. *The San Diego Union-Tribune*, A1 and A11, March 21 (reprinted from *The Washington Post*).

Stacey, Judith (1990), *Brave New Families: Stories of Domestic Upheaval in Late Twentieth Century America* (New York: Basic Books).

Stohmayer, Ulf (1998), The Event of Space: Geographic Allusions in the Phenomenological Tradition. *Environment and Planning D: Society and Space*, 16, 105–21.

Struening, Karen (2001), Fatherless Families: Rousseau and the Contemporary Defense of the Gender-Structured Family. *Women and Politics*, 21(1), 85–106.

Sullivan, Oriel (2004), Changing Gender Practices with the Household: A Theoretical Perspective. *Gender & Society*, 18(2), 207–22.

Swerdlow, Amy, Renate Bridenthal, Joan Kelly and Phyllis Vine (1989), *Families in Flux* (New York: Feminist Press).

Thien, Deborah (2005), After or Beyond Feeling: A Consideration of Affect and Emotion in Geography. *Area*, 37, 450–56.

Thrift, Nigel (1996), *Spatial Formations* (London: Sage).

______ (1999), Steps to an Ecology of Place. In Doreen Massey, John Allen and Philip Sarre (eds) *Human Geography Today*, pp. 295–322 (Cambridge: Polity Press).

Tocqueville, Alexis de (1863), *Democracy in America.* Translated by Henry Reeve (Cambridge, MA: Sever and Francis).

Tongs, R. (1993), *Feminist Thought* (New York and London: Routledge).

Turner, Frederick Jackson (1921), *The Frontier in American History* (New York: Henry Holt and Company).

Varley, Ann (2007), A Place Like This? Stories of Dementia, Home and the Self. *Environment and Planning D: Society and Space*, 26(1), 47–67.

Whitehead, Stephen (2007), Patriarchal Dividend. In M. Flood, J. Kegan-Gardiner, B. Pease and K. Pringle (eds) *International Encyclopedia of Men and Masculinities*, pp. 467–8 (London and New York: Routledge).

Williams, Linda (1991), Film Bodies: Gender, Genre, Excess. *Film Quarterly*, 44(4), 2–13.

Willis, Katie (2005), Latin American Urban Masculinities: Going Beyond "the Macho." In Bettina van Hoven and Kathrin Hörschelmann (eds) *Spaces of Masculinities*, pp. 97–110 (London and New York: Routledge).
Wilson, Elizabeth (1991), *The Sphinx and the City: Urban Life, the Control of Disorder and Women* (Berkeley, CA: University of California Press).
______ (1993), Is Transgression Transgressive? In J. Bristow and A.R. Wilson (eds) *Activating Theory: Lesbian, Gay, Bisexual Politics*, pp. 107–117 (London: Lawrence and Wishart).
Winnicott, D.W. (1971), *Playing and Reality* (London: Tavistock).
Wollen, Peter (1994), The Cosmopolitan Ideal in the Arts. In George Robertson, Melinda Mash, Lisa Tickner, Jon Bird, Bary Curtis and Tim Putman (eds) *Travellers' Tales: Narratives of Home Displacement*, pp. 186–96 (London and New York: Routledge).
Wylie, John (2002), An Essay on Ascending Glastonbury Tor. *Geoforum*, 33, 442–54.
Young, Iris Marion (1990), *Justice and the Politics of Difference* (Princeton, NJ: Princeton University Press).
______ (1997), *Intersecting Voices: Dilemmas of Gender, Political Philosophy and Policy* (Princeton, NJ: Princeton University Press).
Zita, Jacqueline (1998), *Body Talk: Philosophical Reflections on Sex and Gender* (New York: Columbia University Press).

Index